Interactive Technologies and Autism
Second Edition

Synthesis Lectures on Assistive, Rehabilitative, and Health-Preserving Technologies

Editor

Ronald M. Baecker, *University of Toronto*

Advances in medicine allow us to live longer, despite the assaults on our bodies from war, environmental damage, and natural disasters. The result is that many of us survive for years or decades with increasing difficulties in tasks such as seeing, hearing, moving, planning, remembering, and communicating.

This series provides current state-of-the-art overviews of key topics in the burgeoning field of assistive technologies. We take a broad view of this field, giving attention not only to prosthetics that compensate for impaired capabilities, but to methods for rehabilitating or restoring function, as well as protective interventions that enable individuals to be healthy for longer periods of time throughout the lifespan. Our emphasis is in the role of information and communications technologies in prosthetics, rehabilitation, and disease prevention.

Interactive Technologies and Autism, Second Edition
Julie A. Kientz, Gillian R. Hayes, Matthew S. Goodwin, Mirko Gelsomini, and Gregory D. Abowd
2020

Zero-Effort Technologies: Considerations, Challenges, and Use in Health, Wellness, and Rehabiliation, Second Edition
Jennifer Boger, Victoria Young, Jesse Hoey, Tizneem Jiancaro, and Alex Mihailidis
2018

Human Factors in Healthcare: A Field Guide to Continuous Improvement
Avi Parush, Debi Parush, and Roy Ilan
2017

Assistive Technology Design for Intelligence Augmentation
Stefan Carmien
2016

Interactive Technologies and Autism, Second Edition
Julie A. Kientz, Gillian R. Hayes, Matthew S. Goodwin, Mirko Gelsomini, and Gregory D. Abowd

ISBN: 978-3-031-00476-6   print
ISBN: 978-3-031-01604-2   ebook
ISBN: 978-3-031-00038-6   hardcover

DOI 10.1007/978-3-031-01604-2

A Publication in the Springer series
SYNTHESIS LECTURES ON ASSISTIVE, REHABILITATIVE, AND HEALTH-PRESERVING TECHNOLOGIES
Lecture #13
Series Editor: Ronald M. Baecker, University of Toronto

Series ISSN 2162-7258 Print   2162-7266 Electronic

Interactive Technologies and Autism

Second Edition

Julie A. Kientz, University of Washington

Gillian R. Hayes, University of California, Irvine

Matthew S. Goodwin, Northeastern University

Mirko Gelsomini, Politecnico di Milano

Gregory D. Abowd, Georgia Institute of Technology

SYNTHESIS LECTURES ON ASSISTIVE, REHABILITATIVE, AND HEALTH-PRESERVING TECHNOLOGIES #13

ABSTRACT

This book provides an in-depth review of the historical and state-of-the-art use of technology by and for individuals with autism.[1] The design, development, deployment, and evaluation of interactive technologies for use by and with individuals with autism have been rapidly increasing over the last few decades. There is great promise for the use of these technologies to enrich lives, improve the experience of interventions, help with learning, facilitate communication, support data collection, and promote understanding. Emerging technologies in this area also have the potential to enhance assessment and diagnosis of autism, to understand the nature and lived experience of autism, and to help researchers conduct basic and applied research.

The intention of this book is to give readers a comprehensive background for understanding what work has already been completed and its impact as well as what promises and challenges lie ahead. A large majority of existing technologies have been designed for autistic children, there is increased interest in technology's intersection with the lived experiences of autistic adults. By providing a classification scheme and general review, this book can help technology designers, researchers, autistic people, and their advocates better understand how technologies have been successful or unsuccessful, what problems remain open, and where innovations can further address challenges and opportunities for individuals with autism and the variety of stakeholders connected to them.

KEYWORDS

autism, autistic, interactive technologies, technology, computing, human-computer interaction, desktop, web, Internet, video, multimedia, mobile, smartphones, tablets, shared active surfaces, tabletop computing, virtual reality, multi-sensory environments, augmented reality, sensors, wearable computing, robots, robotics, natural user interfaces, natural input, pen input, voice input, gestures, speech, tangible computing, tactile computing, eye tracking

---

[1] In this book, we use "people with autism," "person with autism," "autistic," and other terms interchangeably, because there is no agreed upon standard for terminology by the community. We encourage everyone to use whatever labels they prefer for themselves and to request that others respect those labels. We hope our openness to a variety of terminology is seen here as respectful of that agency and self-determination. See Kenny et al. (2016) for more on terminology preferences.

To Maya, Rohan, and Shwetak—J.A.K.

To Warner, William, and Steve—G.R.H.

To Sage, Sophia, and Emmeline—M.S.G.

To Cettina, Pino, and Ajò—M.G.

To Aidan, Blaise, Mary Catherine, Meghan, Sara, and Richard—G.D.A.

Contents

Figure List

Foreword to the First Edition

This book is a product of pioneering research that the authors, Julie Kientz, Matthew Goodwin, Gillian Hayes, and Gregory Abowd, have conducted—some together in different combinations and some separately—at various research universities such as Georgia Institute of Technology, Massachusetts Institute of Technology, Northeastern University, University of Washington, and University of California, Irvine. The book's immediate impact is two-fold: (1) to demonstrate that interactive technologies are used not only "for," but also "with" and "by" individuals with autism, and thus to acknowledge their agency, autonomy, and creativity; and (2) to demonstrate that interactive technologies are interactive not only in a dyadic user-technology sense, but also that their use mediates interactions within social networks that include individuals with autism as well as other "stakeholders" in their well-being and participation (i.e., family members, peers, teachers, and practitioners).

The authors' comprehensive review of interactive technologies that have been, and are being created and used, takes stock of the state of technology for autism directed at three areas: enriching interventions, facilitating communication, and supporting data collection. This book expects a significant intellectual investment from its audience. The presentation of content is appropriately complex and does not talk down to the readers. Readers interested in benefiting from this volume will have to inhabit the intricate conceptual universe that the authors have built. In addition to being an invaluable resource for individuals with autism and their families, this book will be useful for researchers and practitioners coming into this very important field.

Olga Solomon, Ph.D.
Division of Occupational Science and Occupational Therapy
University of Southern California
2013

Acknowledgments

We would like to thank the following people who provided input directly on the writing of this manuscript, providing reviews, discussions, or comments on the content and writing: Rosa Arriaga, Catalina Cumpanasoiu, Rana el-Kaliouby, Emma Feshbach, Vicki Friedman, Dan Gillete, Joshua Hailpern, Yi Han, Alexis Hiniker, Hwajung Hong, Maithiliee Kunda, Kerri McCanna, Nazneen, Rosalind Picard, Agata Rozga, Kiley Sobel, Katta Spiel, Olga Solomon, Oliver Wilder-Smith, and Miriam Zisook.

In addition, we wish to thank the following people for their influence on our thinking and informing our understanding of the field of interactive technologies for autism in significant ways: Lauren Adamson, Fatima Boujarwah, LouAnn Boyd, Nate Call, Sophia Colamarino, Meg Cramer, Geraldine Dawson, Michelle Dawson, Anind Dey, Joe Donnelly, Theresa Finazzo, Christopher Frauenberger, Mark Harniss, Juane Heflin, Sen Hirano, Lisa Ibañez, Portia Iversen, Lucio Moderato, Ron Oberleitner, Linda O'Neal, David Nguyen, Jim Rehg, Kate Ringland, Ivan Riobo, Carlo Riva, Elisa Rossoni, Clarence Schutt, John Shestack, Andy Shih, Wendy Stone, Khai Truong, Simon Wallace, Ping Wang, Tracy Westeyn, and Michael Yeganyan.

Finally, we wish to acknowledge the following funding sources that supported the authors during the writing of this manuscript: National Science Foundation (Awards 0846063, 0952623, 1029585, 1029679), Simons Foundation for Autism Research Initiative, Nancy Lurie Marks Family Foundation, and the Intel Science and Technology Centers on Social Computing and Pervasive Computing.

CHAPTER 1

Introduction

The use of interactive technologies for, by, and in support of people with autism has grown dramatically over the last few decades. Notably, this research is still dominated by a focus on children; however, some recent scholarship has begun to more fully embrace the wide range of experiences of autistic adults. We attempt in this second edition to more explicitly broaden this review and call out those challenges. Some people claim that autistic people show particular interest in interactive technologies, particularly when they are made inclusive and accessible (Mazurek et al., 2012). During the last two decades, however, use of a variety of personal, social, and organizational technologies has simply become commonplace for all people. This review, then, does not attempt a comprehensive examination of technology use by people with autism and related conditions. Rather, the following pages are dedicated to a consideration of the empirical and published research focused on use of interactive technologies to support and enhance clinical and therapeutic goals for screening and diagnosis, educational goals for intervention monitoring, and other research efforts in support of this community. We take note, when and where possible, of existing gaps we identify in the literature, such as the focus on children noted above or the limited research published by people with autism about their own communities.

This second edition of this book comes nearly a decade after the first version. Our intention then, as now, is to help readers understand some basics of autism and then provide a broad review of the research literature that demonstrates the role technology has served, with some hints to roles it might serve in the future. We note, however, that none of the authors of this book themselves identify as autistic, though we are variably parents, siblings, friends, cousins, and caregivers of people with autism from infant to adult. This book, therefore, is drafted with some level of third-party distance but caring hearts and open minds and a healthy dose of chagrin. In particular, in this second edition, we have made an explicit point to gain feedback from and reviews by people with autism and related conditions.

We expect this book can be useful to new researchers to the area of autism and technology as well as more experienced researchers who are looking to identify new potential areas of focus. We also expect this may be useful to employers, teachers, parents, caregivers, friends, children, and autistic people who want to know more about the research surrounding interactive technology use for autism. We recommend greater discussions with a variety of community groups for those who want to better understand the lived experience of use of these technologies beyond what is represented explicitly in the research literature we cover here. We aim to identify problems that still need to be solved and suggest promising avenues for further development and evaluation, all in service of advancing science, technology, and quality of life for individuals with autism. We also hope to encourage more research in this space, as there are many interactive technologies that are being used for autism that show much promise but have not yet been scientifically validated.

1.1   INTRODUCTION TO AUTISM

Autism was first described as a syndrome by Leo Kanner, a child psychiatrist at John Hopkins University (Kanner, 1943). In his seminal work, Kanner characterized 11 children who shared what he understood to be a fundamental inability to relate to other people, a failure to use language to convey meaning, and an almost obsessive desire to maintain sameness. Our understanding of these "symptoms" has, of course, evolved over time, and we would never recommend using this kind of deficit-based language to describe autism in a modern context. Kanner also noted that anxiety played a prominent role in the clinical presentation of autism; the children he observed often displayed intense fears of common objects. Bettelheim (1967) suggested autism was the result of inadequate nurturing by emotionally cold, rejecting parents, a theory that prevailed until the late 1960s and did a great deal of harm to many families. Rimland (1964) and Rutter (1970) provided persuasive arguments that autism had an organic etiology, the most influential of their findings being that approximately 25% of children with autism developed seizures in adolescence. Current science is in substantial agreement that autism is a complex neurodevelopmental condition with underlying organic genetic and neurological differences, and it is not caused by parenting deficiencies or other social factors. However, some debate continues to exist around epigenetic phenomenon, environmental triggers and contexts, and the cultural creation of types of autism and autistic symptomology. Autism occurs across all socioeconomic levels, in all cultures, and in all racial and ethnic groups (Dyches, Wilder, and Obiakor, 2001). Furthermore, autism is not necessarily debilitating, and autistic people often live happy, full, fruitful lives in a wide variety of contexts. Being autistic, just like being tall or a woman or having brown hair, can be beautiful and wonderful in accepting and accessible societies. It can also be challenging and difficult, particularly for people who experience other co-occurring disabilities or who must live and work in inaccessible contexts.

There is no specific biomarker, laboratory test, or behavioral assessment procedure to identify autism; it is defined exclusively by past and present behavior determined from developmental history interviews (e.g., Autism Diagnostic Interview), parent reporting on current behavior (e.g., Modified Checklist for Autism in Toddlers), and structured and semi-structured tasks that involve social interaction between an examiner and a person being diagnosed (e.g., Autism Diagnostic Observation Schedule). Most people in research studies are children, which can contribute to the idea that all people with autism are children or child-like. This infantilization—even when implicit, explicit, or accidental—can be of great harm to adults with autism and must be addressed in future research. As the increasing number of children diagnosed with autism age into adulthood, and more adults seek diagnoses later in life, this expanded view of autism that incudes highly successful adults must also be considered, something that is a relatively large gap in current research.

Autism is a spectrum condition covering a wide range of ability levels and encompassing a diverse set of symptoms, ranging from more severely affected to what many references as high functioning (often referred to as high-functioning autism or HFA). Despite this heterogeneity, all individuals on the autism spectrum are characterized by qualitative (i.e., exceptional and not merely delayed development) impairments in social-communication and restricted and repetitive interests, activities, and behavior (see DSM-5, American Psychiatric Association, 2013). We note that the DSM-IV (American Psychiatric Association, 2000) included Asperger's Syndrome as part of Autism Spectrum Disorders. The DSM-V, introduced in 2013, has removed that sub-classification. A number of the articles reviewed in this book were designed for or tested with individuals with Asperger's Syndrome as defined by the DSM-IV. The DSM-IV diagnosis also classified other disorders as Autism Spectrum Disorders, such as Childhood Disintegrative Disorder, Rett's Disorder, and Pervasive Developmental Disorders-Not Otherwise Specified, which have also been removed from the DSM-V criteria. There has been some amount of controversy surrounding the new diagnostic criteria, and not everyone agrees. For the purposes of this review, we make use of the diagnostic category in common use at the time the research was conducted. However, outside of a formal research review, we recommend deferring to those titles and labels that autistic people themselves prefer when they are able to state a preference and that their friends and family closest to them prefer when they are not.

Autism is more commonly diagnosed in genetic males than females, with a ratio of 4:1 widely reported across samples (e.g., Fombonne, 2002). Many individuals with autism, but not all, also have moderate to severe cognitive impairments (Fombonne, 1999);[1] suffer from seizures, with onset most often occurring during either the preschool or adolescent years (Volkmar and Nelson, 1990); and/or experience co-occurring neurodevelopmental conditions, for example, Tourette syndrome (Sverd, Montero, and Gurevich, 1993) and attention deficit hyperactivity disorder (ADHD) (Ghaziuddin, Tsai, and Gha- ziuddin, 1992).

Early research suggested that autism was originally considered relatively rare, occurring at a rate of 4 to 6 affected individuals per 10,000 (Lotter, 1967; Wing and Gould, 1979). In the mid-1980s, diagnostic criteria broadened, and rates of 10 per 10,000 were found in total population screenings (Bryson, Clark, and Smith, 1988). More recent studies that focus on preschool children utilize standardized diagnostic measures of established reliability and validity, employ active ascertainment techniques, and yield prevalence estimates of 60–70 per 10,000—translating into approximately 1 in 150 across the spectrum of autism (Baird et al., 2001; Bertrand et al., 2001; Chakrabarti and Fombonne, 2001). At the time of writing, the most recent estimates from the Centers for Disease Control and Prevention (CDC) suggest a rate of 1 in 59 children (Baio et al.,

[1]   For a caveat to these estimates, see Mottron et al. (2006).

2018) but do not state a prevalence rate for adults. Another study reported that 1 in 50 students in U.S. schools receive services related to autism (Blumberg et al., 2013), although no independent replications have confirmed this prevalence estimate. Regardless of the true rate of autism, more than 2 million people in the U.S. are currently believed to carry the diagnosis.

This apparent increase has led to dramatic claims, particularly in the popular media, for an "epidemic" of autism. However, according to reviews by Wing and Potter (2002), Rutter (2005), and Gernsbacher and colleagues (2005), there are several possible reasons for the observed increase in autism rates, including: (1) changes in diagnostic practice; (2) increased awareness among parents, professionals, adults with autism themselves, and the general public of the existence of autism; (3) development of specialist services; (4) differences in methods used in studies; and (5) a possible true increase in numbers.

Individuals with autism often encounter challenges across the entire range of developmental milestones and areas. Of course, development is traditionally thought of as primarily child development, but autism, as a lifelong condition, must consider other aspects of human development and developmental milestones later in life. These might include milestones like entering and exiting child-bearing years, other physical changes surrounding aging, and changes in our socio-emotional development such as through work and interpersonal relationships. The unfolding and maturation of developmental competencies are affected to a greater or lesser degree in autism, depending on each individual's pattern of behaviors, extent of impairment in relation to their environment, and level of support. Autism is almost always a chronic condition; however, the abilities and challenges of individuals is not static. Just as other people do, individuals with autism continue to grow, learn, and develop over the course of their lives. While there is great variability across the autism spectrum, and individual differences make it difficult to generalize across the condition, some general patterns of development can be offered as children with autism move from infancy through adolescence, adulthood, and beyond.

Autism, as with other complex conditions, typically manifests quite differently across people who share the same diagnostic label. Moreover, autism may even present itself differently in the same person across settings and over time. This extreme heterogeneity among the autistic community has led to the common colloquial saying, "If you've seen one person with autism, you've seen one person with autism." However, there are some commonalities, which we describe in the subsequent sections, particularly as they relate access to and use of technology.

1.1.1   INFANT DEVELOPMENT

Infancy is a period of dynamic growth and change. During this time, early symptoms of autism are found to cluster around impairments in early emerging social interaction skills, including attentional functioning, preverbal communication, exploration and play, motor imitation, and attachment. One of the important aspects of attention is the ability to identify and focus on salient elements or features in the environment for further processing (James, 1950). In the first year of life, infants with autism are typically found to visually orient less frequently to people as compared to typical and developmentally delayed infants (Osterling, Dawson, and Munson, 2002). This selective bias is found to persist in the second year and beyond (Dawson et al., 1998). More recent findings suggest this pattern of attention in the first year of life may serve as an effective diagnostic criterion for autism (Klin et al., 2009).

In the first nine months of typical development, and before the development of speech, infants are able to effectively communicate their needs by a variety of means, including reaching for a desired object, fussing, or crying. These communicative attempts are usually directed at the goal itself and not at the person that might be instrumental in attaining the goal. At about nine months, infants begin to direct their communicative attempts toward adults by, for instance, making eye contact while reaching for a distant toy. Along with this change, infants begin to substitute early emerging physical gestures (e.g., an open-hand reach) with conventional gestures such as pointing or showing. Emergence of these behaviors at the end of the first year of life marks the beginnings of intentional communication—communication in which a child is aware that his or her behavior affects a listener (Bates, 1979).

Pre-verbal children with autism typically communicate less frequently (Stone et al., 1997) and use less complex combinations of communicative nonverbal behaviors (Stone et al., 1997). Specifically, two-year-old children with autism are less likely to use eye contact or conventional gestures, such as distal and proximal pointing and showing gestures. They are more likely to manipulate a person's hand using hand-over-hand gestures and are less likely to pair their communicative gestures with eye contact and vocalizations compared to developmentally typical peers (Stone et al., 1997). A disproportionately high number of the communicative behaviors observed in young children with autism are also concerned with requesting objects or actions rather than directing another's attention to an object or event to initiate joint attention (Baron-Cohen, 1989; Bruner, 1975; Mundy and Sigman, 1989). Compared to typically developing peers, most autistic children also develop language later, and their language development is marked by the presence of unusual features (Tager-Flusberg, Lord, and Paul, 2005). For instance, pre-verbal children with autism often show atypical patterns of sound production, including imperfections in well-formed syllables and overproduction of atypical vocalizations such as growling, tongue clicking, and trills (Wetherby, Yonclas, and Bryan, 1989).

Differences in functional and symbolic play in relation to other cognitive skills have also been well documented in autistic preschool children (Sigman and Ruskin, 1999). While play skills continue to develop in the preschool period and can be enhanced through prompts, scaffolding, and modeling, children with autism tend to engage in little spontaneous functional and pretend play at this age (Lewis and Boucher, 1988).

Motor imitation and emulation also play an important role in the emergence of both symbolic and social-cognitive skills (Tomasello, Kruger, and Ratner, 1993). Studies on imitation in autistic preschool children consistently report significant differences in this area (e.g., Rogers, 1999). There has been some speculation that attachment, the affective bond between a child and a parental figure (Ainsworth et al., 1978), is deficient in infants with autism, but there is little evidence to support this claim (Capps, Sigman, and Mundy, 1994). And indeed, such assertions can be considered offensive in much the same way as the early "cold mother" ideas about causes of autism. Other impacts of autism on daily life are detailed in later chapters when relevant to the design of specific technologies.

1.1.2   EARLY CHILDHOOD AND SCHOOL-AGE CHILDREN

The developmental characteristics of autism in infancy tend to persist over time. In school-aged children, these characteristics manifest in specific cognitive, behavioral, communication, and social profiles.

Elementary or primary school years bring challenges associated with changing expectations that accompany increasing physical and behavioral maturity. During the period from ages 6–12, children must transition to new learning environments, develop more sustained contact with new peers and adults, and experience a departure from familiar places and routines. These changes can affect many behaviors in all children, as the child is required to adapt to more complex and demanding social environments, learn more sophisticated skills, communicate at a higher level, and process more information. Such experiences, which are common to all school-aged children, can create substantial barriers for those with autism, who sometimes have developmental delays in multiple domains and/or difficulty adjusting to changes in their environments.

Although there is considerable heterogeneity among children with autism, some generalizations can be made. For example, Wing and colleagues have described three subtypes of social behavior—*aloof, passive*, and *active-but-odd*—that capture many of the manifestations of autism seen in the school-age child (Wing and Attwood, 1987; Wing and Gould, 1979). While these labels are offensive when viewed through today's more inclusive approach, they were groundbreaking in the 1970s. They remain useful categories if not useful labels.

The *aloof* profile is most likely to be described as "classically autistic." These children do not seek, and may actively avoid, contact with others and may become very distressed if forced to do so. Despite having verbal abilities, they do not initiate communication, and much of their time may be occupied with stereotyped or other repetitive interests. These children with autism are noted for their unresponsiveness and lack of attempts to initiate interactions with both peers and adults. They often do not play with other children or demonstrate interest in friendships (Rutter, 1974). Impairments in their ability to use eye gaze and gesture appropriately in social situations lead to frequent communication difficulties. "Aloof" children may also be sufficiently unresponsive, making it very difficult to direct and maintain their attention. They may seem deaf at times, even though they are not. Meltdowns are common in these children, particularly when their routine is disrupted or by other circumstances they cannot control. While individuals with these characteristics are most often seen in the preschool age group, some continue in this manner into later childhood, adolescence, and adulthood. Older individuals with this profile are most likely to have severe cognitive impairments (Seltzer et al., 2004).

The *passive* group includes children who do not actively avoid social contact with others, but who nevertheless lack the spontaneous and intuitive grasp of social interaction achieved in normally developing children. They may accept social approaches of others, but often do not have the skills or interest to respond in a way that other children might read as appropriate. Their communication and play behaviors can be more rigid than typically developing peers and sometimes stereotyped. These individuals tend to test higher in terms of language and motor skills than those in the aloof group. Although "passive" children can be easier to engage and support than those who are aloof, they often still require considerable help relating to peers in the classroom or other situations. Some children with autism who start out displaying the aloof pattern of behavior later have a better fit with the passive group. Thus, presentation as aloof versus passive may depend to some extent on the child's developmental level, and a transition from one category to the other may reflect maturation, effective interventions, and/or accumulation of social experiences.

The *active-but-odd* children are those who are usually described currently as having "higher functioning" autism or may not even identify as having an autism diagnosis at all. They actively seek contact with others, but the form and quality of their social approaches are atypical and may sometimes be read as inappropriate by adults or typically developing peers. These people experience difficulty relating socially to peers, even though they may have considerable language skills and may be interested in communicating with others. Behaviors common to this group include repetitive questioning, inappropriate touching, conversations focused exclusively on the child's own narrow interests, and unexpected facial expressions, postures, and gestures. Their social behavior and communication seem to reflect a view of the social world that is literal and concrete, and they can show limited awareness of the feelings, thoughts, and motives of others.

1.1.3   ADOLESCENCE

The clinical presentation of adolescents with autism has not yet been studied as extensively as younger children with autism, and self-identification has been studied even less. Very few studies have examined whether contextual variables such as parental socio-emotional functioning, place of residence, and educational or intervention history predict later outcomes. However, the limited research suggests that adolescents with autism can follow any of three paths: change their behavior dramatically, experience deterioration in their ability to successfully navigate the social world, or continue a stable maturational course (Seltzer et al., 2004). A key variable mediating these developmental trajectories appears to be intellectual ability. Individuals with autism and intellectual disability (defined by an IQ of less than 70, though it is often difficult for IQ to be measured in autistic people, whether children or adults) have significantly greater difficulties in terms of education, work, living situation, and general independence than those with autism and average intelligence (Howlin et al., 2004). Many parents also report that their intellectually disabled adolescent with autism exhibits significant behavior problems, including resistance to change, compulsions, inappropriate sexual behavior, meltdowns, aggression, and self-injurious behavior (DeMyer, 1979).

Some researchers have begun to examine specific patterns observed in adolesences, such as differences in motivation (e.g., Bos et al., 2019). Likewise, friendship, so important in adolescence for most people, has emerged as an area of focus for autism researchers worldwide (e.g., Chang et al., 2019).

For higher functioning adolescents with autism, academic performance can be at or above grade level; however, organizational and social expectations (e.g., keeping track of multiple assignments and long-term projects; moving quickly between classes; avoiding social taboos) can be overwhelming (Klin and Volkmar, 2000). For those with typical or high intelligence levels who identify as being on the autism spectrum, adolescence may also be a time of heightened loneliness, anxiety, and depression as they recognize the profound nature of their difficulties, their differences from others, and their limited opportunities (Green et al., 2000).

1.1.4   ADULTS WITH AUTISM

Unfortunately, very little is currently known about the ways autism manifests in adulthood and how best to support this older segment of the autism population. A potentially worrisome trend, in fact, is the ending of many longitudinal studies in adolescence rather than adulthood (e.g., Ben-Itzchak and Zachor, 2019; Lin et al., 2019). Lifetime longitudinal studies are incredibly expensive and difficult to undertake, but given the paucity of research on adults, it may be worth the investment. In general, the autistic adult experience is a great area for future research, as the adult population of individuals living with autism will be greatly increasing as the increase in the diagnosis of children increases and these children become adults. Studies that have begun to emerge can focus on issues like anxiety and depression that pervade a lifetime (e.g., Uljarevic et al., 2019, Gillan, 2019) or on specific concerns of adults with autism (e.g. sexuality (Parchomiuk, 2019)). There have been some promising studies with adults in the workplace (e.g., Hendricks, 2010; Hedley et al., 2017; Annabi and Locke, 2019; Wills et al., 2019) as well as increasing initiatives in a variety of corporations and non-profits to develop neurodiverse workforces. Likewise, increasing efforts have sought to include autistic adults in the design process, such as the Clap novel tactile anxiety management program (Simm et al., 2014).

1.1.5   THE ROLE OF SOCIAL ENVIRONMENT

A transactional model of development recognizes the individual as an active participant in the developmental process who, through continued and varied interactions with the environment, comes to adapt, learn, and develop (Bronfenbrener, 1979). In addition to the individual-level characteristics of autism described above, there are also factors relating to the family and school environment that contribute to developmental outcomes in this population. The following briefly reviews the roles that families and schools play in autism.

Questions about interactions between family context and the developmental trajectory of autism are relatively understudied. Studies in autism have focused primarily on child variables and child outcomes or on the stress of life for parents. Family variables, considered to be critical to general early intervention research (such as socioeconomic status, stress, supports available, and parents' involvement in a child's development), have not been well addressed in outcome studies (Gresham and MacMillan, 1997). However, there are a few studies showing that family involvement in intervention is a strong predictor of outcome in children with autism (Dawson and Osterling, 1997; Dunlap, 1999; Lord and McGee, 2001; Rogers, 1998). Family influences on adults have not been covered in the research literature and are an important area of future research. Conversely, borrowing from the broader developmental literature, negative family factors including limited financial resources, lack of appropriate services, and insufficient support systems most likely produce unfavorable prognoses in children with autism. Environmental risk factors, such as lack of services and negative attitudes toward disability, probably also negatively influence the development of a child with autism as well as concerns of adults, such as barriers to employment and post-secondary education. However, as mentioned previously, there is very little systematic investigation of these factors in autism, particularly beyond early childhood.

Research also suggests that school-based interventions that utilize a structured environment and intensive early behavioral, language, and social skills training (Rogers, 2000; Smith, Groen, and Wynn, 2000) can be effective in helping children with autism. Structure helps students with autism by making elements of the learning environment clearer and more predictable (Olley, 2005). Individualized interventions and education goals that are developmentally appropriate have also been shown to address this population's abilities and impairments and facilitate development (Schreibman and Ingersoll, 2005). These same kind of highly structured approaches have begun to be explored for adults who identify as autistic as well in learning (e.g., Duggal et al., 2019) and workplace contexts (e.g., Wills et al., 2019).

1.1.6   MOVING TOWARD ACCEPTANCE AND INCLUSION

The medical model of disability most often invoked—sometimes unintentionally—in the work we review focuses on physical and functional limitations a person might exhibit, and thereby looks to augment, assist, or adjust for these deficiencies. Other approaches include a social model that focuses on condition management and independent living, rather than "fixing" a person with a disability (Zola, 1983), as well as post-modern approaches that privilege each unique individual's lived experience (Pinder, 1996). There is increasing movement toward these models of acceptance and inclusion that both address the unique needs of individuals with autism as well as the social and structural barriers present in their lived experience. Research has also focused on describing the relationship between assistive technologies and disability studies (Mankoff et al., 2010), as well as consideration of neurodiversity in human-computer interaction research (Dalton, 2013). One promising area of future research is to identify how some of the technologies reviewed in this article might be re-envisioned to support more of a social acceptance and inclusive perspective rather than one focusing on rehabilitation, such as pioneering work in the education space on inclusive education (Odom and Diamond, 1998), building on pioneering work on re-visioning technologies and design methods to focus on inclusion and engage directly with individuals with autism (e.g., Sobel et al., 2015; Spiel et al., 2017). Similarly, Milton (2012) describes the "double empathy problem" in an attempt to move away from autism being defined as a deficit of some sort, frequently in theory of mind, but instead being a difference in reciprocity and mutuality.

1.1.7   ADDITIONAL CHALLENGES WITH AUTISM

Research evidence in the behavioral, educational, and social sciences indicates that early diagnosis and intervention can be essential to achieving greater independence. Over the last two decades, popular press and parenting forums have taken this advice to heart, pushing parents and teachers to act more and more quickly. Thus, caregivers can feel that they are in a race against time to find interventions that work for their child. However, many interventions may or may not work for any particular child, and these interventions are often applied simultaneously. Interventions commonly take the form of pharmacology, special diets, occupational therapy and sensory integration, behavioral therapies such as applied behavior analysis or functional behavior analysis, and symptom-specific support such as speech or language therapy. It may also be the case that the predominant push to conduct research with autistic children rather than adults has been created in part by this emphasis on "early intervention."

The difficulty obtaining an early diagnosis, inherent heterogeneity in clinical presentation, and lack of a complete evidence-based research on the effectiveness of interventions all present significant challenges to those who support individuals with autism. While the condition has been known in the literature since the 1940s, we are increasingly learning how complex a condition autism is and how its etiology is at best a complex combination of genetic predisposition and environmental hazards. Despite our incomplete knowledge, there is still a lot we do understand about the challenges of autism in everyday life, and hence a lot that can be done to address them.

1.2   COMPUTER USE BY INDIVIDUALS WITH AUTISM

It is widely accepted that computing applications in multiple domains are largely successful when used by people who identify as autistic. Compared to the accessibility challenges seen with physical disabilities (e.g., blindness or paralysis) or with other cognitive differences (e.g., related to aging and dementia), autism is somewhat unique in that most computing interfaces mostly work for autistic people. The earliest articles we are aware of that discuss the use of computers and why they are promising for individuals with autism is that of Colby and colleagues (Colby, 1973; Colby and Smith, 1971). Another early work was the use of a microprocessor-based system for training individuals with autism using prompting and data collection (Rathkey et al., 1979). Panyan (1984) suggested that computer use in autism could increase learning rates, ability to work independently, creativity, and attention and social behaviors. More recently, Silver and Oakes (2001) outlined several factors that may help explain the particular affinity for computers observed in individuals with autism. First, individuals with autism often have difficulty filtering sensory information that is not salient to their daily interactions (Rutter and Schopler, 1987). Computer screens allow information to be abstracted or limited to only relevant information, thereby supporting the filtering process. Second, many individuals with autism are often confused by unpredictability, social nuance, and rapid changes present in the non-computerized physical world. Computers are much more predictable than humans and do not require social interactions. Additionally, computational interactions can be repeated indefinitely until the individual achieves mastery. Third, computers can provide routines that are explicit, have clear expectations, and deliver consistent rewards or consequences for responses, which can encourage engagement with educational and assistive technologies by allowing an individual to make choices and take control over their pace of learning. Fourth, content can be selected and matched to an individual's cognitive ability and made relevant to their current environment, and photos can be used to help generalize to the real world. Finally, learning experiences can be broken down into small and logical steps and progress at a rate necessary for conditioned reinforcement. The data collected by computers can also be useful for assessing progress in learning, self-monitoring, and adaptive uses to support employment, greater autonomy, and postsecondary education.

In general, due to the individualistic nature of the autism experience, computer-based interventions can be tailored to an individual's needs or even special interests (Morris et al., 2010), which can potentially enhance learning and maintain interest over time. Because of these perceived benefits of using computers, they have become an integral part of a number of interventions and educational programs. They have also become a good way of supplementing face-to-face therapies that are time, cost, and/or other-resource prohibitive.

A number of approaches to designing interactive technologies for individuals with autism have been proposed. Lahm (1996) conducted a study assessing software features and what engages individuals with autism in the classroom and found that technology with higher interaction requirements and those that use animation, sound, and voice were more likely to captivate attention. Several papers discuss the experience of using participatory design with children with special needs (Benton et al. 2012; Frauenberger et al., 2011, 2012; Constantin et al., 2019) and more specifically children (Millen et al., 2011; Ghanouni et al., 2019b) and adolescents with autism (Madsen et al., 2009a; Zhu et al., 2019). Navedo et al. (2019) explicitly called out a strengths-based approach for such design. Kaufman and colleagues (2011) describe their experience with iterative design for advancing the science of autism, and Porayska-Pomsta and colleagues (2012) discussed an interdisciplinary approach to technology design for individuals with autism. Finally, Putnam and Chong (2008) conducted a survey with parents of individuals with autism to understand what needs they have in computing software and found that social skills, academic skills, and organization skills were the most important areas for interactive technologies.

1.3   OTHER REVIEW ARTICLES

This book is not the first to review the space of autism and technology. In our literature review, we encountered a number of summary articles, ranging from systematic and meta-analytic reviews (e.g., Ramdoss et al., 2011; Grynszpan et al., 2013) to discussion or thought pieces about the future of technology in support of autism (e.g., Goldsmith and LeBlanc, 2004). We also encountered many articles that sought to review specific types of technologies for use with individuals with autism, such as robotics (e.g., Feil-Seifer and Matarić, 2009), virtual reality (e.g., Parsons and Mitchell, 2002), tabletops (e.g., Piper et al., 2006), video instruction (e.g., Ayres a nd Langone, 2005a), pervasive computing (e.g., Kientz et al., 2007), speech output (e.g., Schlosser and Blischak, 2001), and computer-mediated communication (e.g., Burke et al., 2010). In addition, we encountered a number of articles that describe the use of technologies for specific purposes, such as communication (e.g., Mirenda, 2001), functional skills instruction (e.g., Ayres and Langone, 2005b), and social skills (e.g., Reed et al., 2011; Wainer and Ingersoll, 2011). There have also been review articles that discuss the role of technology in supporting family and caregivers of individuals with autism (e.g., Oberleitner et al., 2006; Solomon, 2012). Spiel et al. (2019) brought important voice to these discussions with their critical literature review in 2019, including critiques directed at the first edition of this book, which we have taken to heart and incorporated as much as possible. Finally, important discussions are taking place about mapping a research agenda for autism and technology research that incorporates all stakeholders (Parsons et al., 2020). They propose a framework for future research that focuses on social inclusion, perspectives, participation, and agency, suggesting that research focusing on autism as a "disorder" to be addressed is exclusionary and that an agenda promoting diversity, inclusion, and equity and engages the voices of autistic individuals is the future of research in this space.

In this book, we make no attempt to include every article written on the subject of technology and autism. Indeed, over the course of writing this book, and in the 10 years between the first and second editions, many more relevant articles came to our attention, a testament to the promise and interest in this rapidly growing area. Thus, we focus here on providing a classification scheme from which to overview the general space that is beneficial to those seeking a basic introduction to the area, while providing enough detail to foster future research and allow current researchers to position their work among existing literature. We encourage readers to characterize their own work using our provided classification scheme, discussed in Chapter 2. In the current review, we also largely leave the kind of critique of technologies that one might invoke based on different models of disability to the side in favor of a broader summarization of technological approaches.

1.4   STRUCTURE OF THIS REVIEW

The structure of this review is as follows. We first begin with a discussion of our method for identifying published papers included in this review and provide a description of the classification scheme we developed for organizing different technologies according to input and output types, domain, goal, target end user, setting, publication venue, empirical support, technological maturity, and engagement of individuals with autism. The next seven chapters constitute the core of our review and are based on seven general types of interactive technology platforms, including personal computers and use of the Web, mobile devices, shared active surfaces, virtual and augmented reality, sensor and wearable technologies, robotics, and natural user interfaces. In each core review chapter, we review technologies that use the platform and provide a discussion of challenges and future directions for research using that platform. We conclude the book with some overall discussion points and thoughts for the future of interactive technologies for autism.

CHAPTER 2

Methods and Classification Scheme

In this chapter, we provide a description of the methods we used to identify and classify interactive technology research included in this review. Through a high-level analysis of the existing literature, we developed a classification scheme to help categorize each technology approach. Several frameworks could be developed around the same body of literature, and in fact, we experimented with multiple approaches while drafting this content for the first edition. Ultimately, we settled on an approach that is both descriptive and explanatory, while supporting the potential for exploration and innovation going forward. This classification scheme has also evolved over the years as new technology platforms and applications are explored. For the second edition, for example, virtual reality had greatly expanded in the intervening decade.

2.1 METHODS

This section is intended to provide an overview of the use of interactive technologies by and for autistic people—including those with childhood diagnoses, those with adult diagnoses, and those who may be undiagnosed but nevertheless identify as being on the autism spectrum--as well as for researchers new to the area, who may already be experts in a variety of social, medical, and computer science fields, or who may be new to research altogether. Given the rapid growth rate in this field, we did not set out to conduct a comprehensive review of the literature. Rather, with a focus on being as inclusive as possible, we set out to understand both the research and design spaces of this important and continually growing field. Notably, there are a wide variety of relevant commercial products in this space as well. Given our focus on research and the difficulty of conducting any type of comprehensive review of commercially available products, we limit our discussion primarily to research projects and research-validated products but do include commercial products when they are particularly relevant. As a result of this approach, we did not have specific inclusion and exclusion criteria for works included in this review. Others have conducted systematic reviews and meta-analyses of the autism and technology literature that we defer to for this level of analysis (Ramdoss et al., 2011; Grynszpan et al., 2013).

In gathering papers to include in this review, we conducted searches on the ACM Digital Library, IEEE Xplore, PubMed, ERIC, and PsychInfo. We also searched abstracts of the International Society for Autism Research (INSAR) from the last five years on Google Scholar to identify published papers from those projects. Keywords included "Autism," "Asperger," "PDD-NOS," "Technology," "Computer," "Robot," "Sensor," "Virtual Reality," and "Mobile." We then searched citations of the resulting articles for additional papers to include. From the resulting papers, we narrowed down the list to those that fit our definition of interactive technologies and included the most recent articles for each application identified. We included technologies that ranged from demos to fully functional or publicly available technologies and those that have been used by, or specifically designed for, individuals with autism and their caregivers. We did not include technologies that had the potential to be used for autism but had not yet been applied to this domain. We note that there may be additional and related search terms beyond those identified, and while we tried to be as inclusive as possible, the search we conducted was not intended to be systematic. The review we conducted was first done in 2013 when the first version of this book was published, and then again in 2019 for the second edition.

Our review and classification scheme are based on technologies described specifically *in the papers we identified*, as opposed to hypothetical or extrapolated uses, such as those mentioned in a critical review intro or discussed in conclusion/future directions section or based on what we know from outside knowledge or future work. We focused our search on those technologies that originated from the research community or have a basis in research-validated intervention techniques. This ended up excluding a number of applications from popular media, such as games for children with autism found in the Apple App Store[2] or on Google Play.[3] These marketplaces are rich with different applications, but in general, they are beyond the scope of this book unless they have been studied in the scientific literature.

2.2   CLASSIFICATION SCHEME

To organize the papers, we conducted bottom-up coding of different aspects of 26 influential papers across a broad spectrum in the area of technology and autism to determine a set of characteristics that define their use, listed in Figure 2.1. To refine the codes further, we individually applied these codes to a larger set of papers, and then met to discuss the application of the codes and finalize the set. Once there was strong agreement among the authors, we categorized the codes and wrote definitions to develop a classification scheme to help organize existing literature and projects relating to interactive technologies for autism and to help identify areas for future work. This process was repeated and refined for the second edition of this book.

The first edition of this book focused its bottom-up review on 20 papers, and the revision of this book removed some papers and added others that were more relevant, for a total of 26. The original classification scheme from the first edition of the book consisted of eight dimensions, and the final scheme for this latest edition consists of 10 different dimensions along which projects can be categorized. In particular, we decided that the "technology platform" classification scheme did

---

2   http://www.apple.com/osx/apps/app-store.html
3   http://play.google.com/store

not reflect the changing nature of technology as well, and thus we replaced it with input and output modalities and only used technology platform as an organizing scheme for chapters. In addition, the shift of research in this space from being designed in service of people with autism to be more about designing with autistic people necessitated the inclusion of a classification scheme for how people with autism were included in the technology design process. This shift hopefully provides additional considerations for the agency of all people with autism, though particularly for autistic adults and those who identify as being on the autism spectrum. Finally, we made minor updates within some of the other dimensions to reflect changes in terminology or to collapse or separate sub-dimensions.

Within each dimension, we determined several labels that could describe the work either based on prior literature or on our own initial coding. Below, we list the 11 dimensions, along with associated labels within those dimensions and their operational definitions. For each technology, it is possible that several labels exist within each dimension, such as a technology being used for both home and school settings or one that is used by both individuals with autism and their family members.

2.2.1   INPUT MODALITY

One of the ways that interactive technologies are defined is by their input and output modalities. This is the channel by which interactions from a person are transmitted to a computing system (Karray et al., 2008). Systems might be considered unimodal or multimodal if they engage one or more input types respectively. This section describes the different *input* types that interactive technology may have, meaning how the user gives information to the technology. Input may be captured via a number of methods, including direct connections to the computing device, sensors, computer vision, microphones, etc.

- *Audio*: Any speech or non-speech audio-based input, such as spoken voice dictation, natural speech processing, audio-command input, or non-speech audio input such as humming or snapping fingers.

- *Physical*: Includes any methods that require the user to physically touch something, including touchscreens, keyboards, pointing devices (e.g., mouse, trackball, etc.), video game controllers, buttons, switches, etc.

- *Spatial*: Any input based on movement within a space, including based on gestures, body movement (e.g., accelerometers or eye tracking), orientation, or location-based tracking such as with GPS or indoor localization.

2.2.2   OUTPUT MODALITY

Output modalities are ways in which a computing system provides information that can be perceived by the user. These are largely based on the five senses of humans and what they are able to perceive.

- *Audio*: Any speech or non-speech output perceived through the ears, including spoken sound, music, sound effects, or other noises.

- *Tactile*: Any output that can be perceived via the user's sense of touch, including haptics and vibration.

- *Visual*: Anything coming into the user's eyesight, primarily through screen-based displays, but may also include projectors, lights, physical object movement (e.g., a robot's eyes blinking), or e-ink displays.

- *Other*: Although less common than the other three, there are other ways that computing devices can provide output including gustation (taste), olfaction (smell), thermoception (heat), nociception (pain), or equilibrioception (balance).

2.2.3   DOMAIN

This category refers to the focus area relevant to autism that the technology targets, such as helping with acquisition of certain skills or addressing certain challenges.

- *Academic Skills*: Includes applications that focus on skills traditionally taught in educational institutions, including math, science, letters, shapes, colors, etc. Language skills would be an academic skill, but because they are often a primary focus for other applications, we included them in their own category. With expanded attention to adolescents and adults, this category also includes a push towards postsecondary education. However, employment-related technologies, are left in the vocational skills section described below.

- *Behavior*: Includes applications that focus on promoting positive behaviors and reducing negative behaviors. This includes repetitive or circumscribed behaviors, interests, or play. May also include both high-level cognitive behaviors and low-level behaviors, such as manipulation of body or objects.

- *Cognition*: Includes applications that support improvement of cognitive functioning, including memory, attention, and information processing.

- *Language/Communication*: Includes applications or projects that focus on learning vocabulary, language acquisition skills, reading, spoken language for communicative purposes, semantics, syntax, morphology, or prosody.

- *Life/Vocational Skills*: Includes skills that allow autistic people autonomy while conducting daily tasks in home, work, or social environments. Includes skills such as clothing, toileting, meal times, time management, transportation, safety, scheduling, and workplace skills.

- *Play*: Includes applications that promote play to engage in learning and play-based activities. Can include both structured and unstructured play activities.

- *Responding/Motor*: Includes applications that focus on an individual's response to stimuli, as well as movement, including fine motor, gross motor, motor fluency, posture, and gestures.

- *Sensory/Physiological*: Includes applications that focus on an individual's sensory or physiological responding, such as perception, activation, recovery, or regulation.

- *Social/Emotional Skills*: Includes applications or projects that focus on emotion recognition, prosocial behaviors (e.g., turn-taking, sharing, etc.), nuances, and figures of speech. In some cases, this consideration includes non-autistic people adapting to and supporting autistic sociality.

2.2.4   GOAL

This category refers to the primary goal of the technology itself. Some technologies related to autism are intended to screen or diagnose, whereas others are intended for interventions.

- *Functional Assessment*: Includes applications or projects focused on the collection and review of data over time to assess an individual's learning, capability, or level of functioning. The data collected is intended for end users and/or people caring directly for individuals with autism.

- *Diagnosis/Screening*: Includes applications that assess the risk of an autism diagnosis in the general population, or that assist in helping make or understand the severity of an autism diagnosis.

- *Intervention/Education*: Includes applications that attempt to improve or produce a specific outcome in an individual with autism. May focus on teaching new skills, maintaining or practicing skills, or changing behaviors.

- *Scientific Assessment*: Includes applications or projects that use technology in the collection and analysis of data by researchers to understand more about autism and its features or characteristics.

- *Parent/Clinical Training*: Includes applications that provide support for caregivers, educators, clinicians, and other professionals to further their own learning and education or improve skills.

2.2.5   TARGET END USER

This category focuses on the person or persons who actually interact with the technology itself and are considered the primary users. It does not include secondary stakeholders or those who may benefit from the technology but do not actually interact with or use it.

- *Person with Autism*: Includes autistic infants, youth, adolescents, and adults. Diagnosis can be anywhere on the autism spectrum or self-identification without a formal diagnosis.

- *Family/Caregiver*: Includes anyone who cares for or supports anyone with an autism diagnosis or who self-identifies as autistic. May include parents, siblings, other family, friends, volunteers, group home staff, etc.

- *Peer*: Can be an adult or child who is a peer to autistic people. Includes both neurotypical individuals as well as those with autism or any other disabilities or chronic conditions. Recognizing the potential overlap with the caregiver category, this group includes people who typically do not perform explicit caregiving activities, such as co-workers. However, of course, as is true for all people, our peers often provide care informally and on an ad hoc basis even when that is not how they primarily identify in their relationship.

- *Professional*: A paid professional who works with autistic people. May include medical professionals, doctors, occupational therapists, physical therapists, speech therapists, applied behavior therapists, or other allied health professionals. Also includes educators and community support providers, such as those who teach or are otherwise involved in the education of students with autism in schools (public or private), including teachers, administrators, school staff, etc. and those who support autistic people in overcoming barriers to postsecondary education and employment, such as job coaches or transition specialists.

- *Researcher*: Anyone intending to collect data or conduct studies about individuals with autism and publish something generalizable about obtained data.

2.2.6   SETTING

The care of individuals with autism takes place in a number of different settings. This category refers to the settings or locations in which the technology is intended for use.

- *Clinic*: A place of professional practice that is not intended for education, such as a doctor's office, therapist's office, or a specialty service provider.

- *Community/Workplace*: Technology is intended for use while the person is in public places like parks, stores, restaurants, etc. and/or places of employment.

- *Home*: An autistic individual's personal and/or shared living space.

- *Research Lab*: Technology is intended for use in a research laboratory under careful observation or for controlled settings.

- *School*: A public or private place for educating individuals with autism. Includes both schools that specialize in autism education as well as general, inclusive classrooms. Could be at all levels from preschool through postsecondary education.

2.2.7   PUBLICATION VENUE

Technology for autism is inherently interdisciplinary, and these disciplines have large variations in expertise and research approaches. This category describes the field from which the research publication originated.

- *Autism-Specific*: Journals or publication venues specifically relating to understanding autism. Examples: *Autism, International Society for Autism Research* (INSAR), *International Meeting for Autism Research* (IMFAR), *Journal of Autism and Developmental Disorders* (JADD), and *Focus on Autism*. There is a conference series specific to autism and technology, *International Conference on Innovative Technologies for Autism Spectrum Disorders*,[4] first held in 2012.

- *Computing*: Journals, conference publications, and other publication venues relating to the fields of computing, computer science, or human-computer interaction. Often included in the Association of Computing Machinery (ACM) or Institute of Electrical and Electronics Engineers (IEEE) digital libraries. This field traditionally publishes in

[4]   http://www.itasd.org/

conference venues in addition to journals. Examples: *Conference on Human Factors in Computing Systems* (CHI), *Ubiquitous Computing* (UbiComp), *Interactive Mobile Wearable and Ubiquitous Technologies* (IMWUT), *Computer Supported Cooperative Work and Social Computing* (CSCW), *Transactions on Human-Computer Interaction* (ToCHI), *Personal and Ubiquitous Computing* (PUC), *Interaction Design and Children* (IDC), *Computers and Accessibility* (ASSETS).

- *Education*: Journal articles or publications focusing on education or special education. Often included in the ERIC digital library. Examples: *American Journal on Intellectual and Developmental Disabilities, Mental Retardation*, and *Journal of Mental Health Research in Intellectual Disabilities*.

- *Medical*: Journal articles or publications from the medical field, including health informatics or general mental health. Often included in the PubMed digital library. Examples: *Journal of the American Medical Association* (JAMA), *Journal of the American Medical Informatics Association* (JAMIA and AMIA conference), and *Journal of Mental Health*.

- *Social/Behavioral Science*: Journals or publication venues from areas in psychology, human development, or sociology. Examples: *Journal of Consulting and Clinical Psychology, Child Development*, and *Behavior Research Methods*.

2.2.8   EMPIRICAL SUPPORT

Many technologies that have been designed related to autism are experimental and may not be scientifically proven yet. To help readers distinguish between these types of technologies, this category describes the type of study that has been completed with the technology in terms of its feasibility, usability, acceptability, efficacy, and effectiveness. We note that in the field of human-computer interaction, smaller N studies are common due to the cumbersome nature of doing rigorous evaluations of novel and often non-robust technology prototypes, and thus feasibility studies are more common. Meanwhile in transition specialties and special education, as well as increasingly in health sciences, "N of 1" studies are increasingly acceptable. In addition, we note that the three levels within this classification are fairly broad. Within a given category, there may be varying levels of quality in terms of study design, number of participants, and level of control.

- *Descriptive*: Study design seeks to observe natural behaviors without affecting them in any way. Common approaches include observational methods (e.g., ethnography), case study methods, and survey methods.

- *Correlational/Quasi-Experimental*: Study design involves comparing groups, without any random pre-selection processes, on the same variable(s) to assess group similarities/differences and/or determine the degree to which variables tend to co-occur or are related to each other. They are similar to experimental study designs but lack random assignment of study participants. Common approaches include nonequivalent groups design, regression-discontinuity design, retrospective designs, and prospective designs.

- *Experimental*: Study designs seek to determine whether a program or intervention had the intended causal effect on study participants. There are three key components of an experimental study design: (1) pre-post test design, (2) a treatment group and a control group, and (3) random assignment of study participants. Common approaches include randomized controlled trials, Solomon four-group design, within-subject design, repeated measures design, and counterbalanced measures design.

2.2.9   TECHNOLOGY MATURITY

This category describes the maturity of the technology used with individuals with autism and its readiness for use or distribution by the general public.

- *Design Concept/Prototype*: The technology is not yet functional. It may be an idea expressed as a sketch, storyboard, interface mockup, etc. May also include non-functional but interactive prototypes such as paper prototypes, Wizard-of-Oz prototypes, video prototypes, demos, etc.

- *Functional Prototype*: A functional prototype has been developed and interacted with the intended users for the target purposes. It has been built for use by the developers to answer specific questions, but may require assistance with setup, use, or maintenance.

- *Publicly Available*: The technology is mature enough that it can be used without assistance from the developers or research team. This might be a commercial product, software that is open source, or applications available for download on websites or on mobile marketplaces. At times, the product was commercially available at the time of review and is not any longer. Commercial enterprises fail. However, in this case, we still categorize the product as being commercially available, because it was so at the time of the original research.

2.2.10   INVOLVEMENT OF INDIVIDUALS WITH AUTISM

It is important to recognize that technology designs are more likely to be successful when the people for whom the technology is designed are included in the design process. To this end, we wanted to recognize and establish the importance of including individuals with autism in the design of technology tools. Druin (2002) pioneered a framework for assessing how children can be involved in the design of technology, ranging from "as users" to "as design partners." Although this framework was initially intended for inclusion of children in the design process, we believe this framework applies similarly to other marginalized and potentially vulnerable populations. For each of the dimensions, we describe whether and how the autistic individual's inclusion was incorporated into the design of the tool.

- *As Users*: Individuals with autism are primarily using technology not designed with their involvement. They may have participated in studies where they are observed using the technology and the results of their use may be published, but the technology was not knowingly used to affect the design of the system.

- *As Testers*: Individuals with autism were involved as testers of the system, and the results of the testing of the system were used to modify and/or improve the system's design. This could happen early in the process through using low-fidelity prototypes and/or later in the process with fully functional systems.

- *As Informants*: Autistic individuals were involved early in the design process and may have contributed some of the initial ideas through participatory design workshops where they help ideate or protype new ideas.

- *As Design Partners*: Individuals with autism were involved through the entire process and were made to be as equal of partners as possible, with special consideration taken to build relationships between the individuals and the design team.

2.3   SUMMARY OF CLASSIFICATION SCHEME

As an example of how we applied the classification, Abaris (Kientz et al., 2005; Kientz et al., 2006), an application that uses digital pen and paper input, speech recognition, and video recording and playback for supporting therapists conducting discrete trial training therapy, would be categorized as:

- **Input Modality** *of Audio and Physical*

- **Output Modality** *of Audio and Visual*

- **Domains** *of Academic Skills, Life/Vocational Skills, and Language/Communication Skills*

- **Goals** *of Functional Assessment and Intervention/Education*

- **Target End Users** *of Professionals*

- **Settings** *of Home and School*

- **Publication Venue** *of Computing*

- **Empirical Support** of Descriptive Study

- **Maturity** *of Functional Prototype*

- **Involvement of Individuals with Autism** *of as Tester*

Figure 2.1 includes 26 papers on technologies reviewed for this book and their associated coding within the classification scheme as a demonstration of the coding scheme. We chose the 26 papers based on their diversity across specific areas within the classification as a way of defining, refining, and testing our scheme's components. This list of papers was expanded and updated in the second revision of this book.

In the subsequent chapters, we use this scheme to describe the different types of interactive technologies. The chapters are organized based on groupings of technology platforms and specific differentiators that are most relevant to the systems, similar to what has been used by other systematic review articles (e.g., Grynszpan et al., 2013). The subsequent chapters include Personal Computers (Chapter 3), Mobile Technologies (Chapter 4), Shared Interactive Surfaces (Chapter 5), Virtual and Augmented Reality and Multi-Sensory Environments (Chapter 6), Sensor-Based and Wearable (Chapter 7), Natural User Interfaces (Chapter 8), and Robotics (Chapter 9). Some applications and technologies might fit into more than one of these categories (such as both mobile devices and a shared interactive surface). We discuss those as part of each chapter and co-reference where appropriate.

Within each chapter, we have included a section that discusses exemplary technology platforms from among the 26 papers used to develop the classification scheme. We describe how these technologies fit within the rest of the scheme beyond the technology platform and then provide a discussion of the overall trends we saw for the given platform. We also discuss opportunities for future work and identify areas that may be of interest to new researchers.

We note that this review provides a snapshot of the current landscape of interactive technologies for autism, and that technology is always evolving and changing, as is our knowledge about autism and cultural expectations about language, inclusion, and diversity. We expect that there will be many new technologies identified beyond those discussed in this review. To keep the community up to date, we have started a public, shared Mendeley.com group called "Interactive Technologies for Autism" that contains all of the references cited in both this book and the previous edition.[5] We welcome the community to add additional references, comment on existing ones, and add tags to references to classify them in our scheme. We also believe the classification scheme may evolve as technologies evolve, and we welcome a discussion of this through the Mendeley group.

---

[5] http://www.mendeley.com/groups/3745371/interactive-technologies-for-autism/

Column groups: **Chapter** (3. PC and Multimedia, 4. Mobile, 5. Shared Active Surfaces, 6. Virtual and Aug. Reality, 7. Sensor & Wearables, 8. Natural User Interfaces, 9. Robotics) · **Input Modality** (Audio, Spatial, Physical) · **Output Modality** (Audio, Tactile, Visual, Other) · **Domain** (Academic, Behavior, Cognition, Language/Communication, Life/Vocational, Play, Sensory/Physiological, Social/Emotional) · **Goal** (Functional Assessment, Diagnosis/Screening, Intervention/Education, Scientific Assessment, Parent/Clinical Training)

Paper	PC & Multimedia	Mobile	Shared Active Surfaces	Virtual & Aug. Reality	Sensor & Wearables	Natural User Interfaces	Robotics	In: Audio	In: Spatial	In: Physical	Out: Audio	Out: Tactile	Out: Visual	Out: Other	Academic	Behavior	Cognition	Language/Communication	Life/Vocational	Play	Sensory/Physiological	Social/Emotional	Functional Assessment	Diagnosis/Screening	Intervention/Education	Scientific Assessment	Parent/Clinical Training
Bauminger et al., 2007		•								•	•		•									•		•			
Bonarini et al., 2016							•	•		•	•	•	•		•		•	•	•	•		•					•
Bondioli et al., 2019			•							•	•		•		•				•	•		•	•				•
Dziobek et al., 2006	•									•	•		•									•					•
Escobedo et al., 2012	•									•	•		•									•			•		
Feil-Seifer et al., 2009							•	•	•	•	•	•	•									•			•	•	•
Fisicaro et al., 2019							•	•	•		•		•	•				•				•					
Garzotto et al., 2017							•	•		•	•		•		•			•	•	•	•	•			•	•	
Garzotto et al., 2018			•							•	•	•	•	•	•	•			•	•	•	•			•	•	
Gelsomini et al., 2018			•							•	•		•	•	•				•	•	•	•			•	•	
Goodwin et al., 2006				•						•	•		•								•				•	•	
Goodwin et al., 2019				•						•	•		•		•						•				•	•	
Hailpern et al., 2009						•		•		•	•		•		•	•						•			•		
Hayes et al., 2008	•									•	•		•		•	•						•	•		•		
Heathers et al., 2019				•						•	•		•								•					•	
Hirano et al., 2010		•								•	•		•		•				•			•			•		
Hong et al., 2012			•					•		•	•		•						•			•			•		
Kientz et al., 2005					•			•		•	•		•					•	•	•			•			•	
Klin et al., 2002					•			•					•									•		•			
Piper et al., 2006		•								•	•		•									•			•		
Sobel et al., 2016		•								•	•		•									•			•		
Suh et al., 2016	•												•			•	•	•									•
Tam et al., 2017			•							•	•	•	•				•			•			•		•		
Tartaro et al., 2008		•						•		•	•		•					•				•				•	
Wallace et al., 2010		•						•		•	•		•									•				•	
Whalen et al., 2010	•									•	•		•		•			•							•		

Figure 2.1: Coding of 26 papers used to define, refine, and test our classification scheme (continued on following page).

Paper	Person with Autism	Family/Caregiver	Peer	Professional	Researcher	Community/Workplace	Clinic	Home	Research Lab	School	User	Tester	Informant	Design Partner	Autism-Specific	Education	Computing	Medical	Social/Behavioral Science	Correlational	Descriptive	Experimental	Design Concept	Functional Prototype	Publicly Available
Bauminger et al., 2007	◆		◆	◆						◆	◆						◆					◆		◆	
Bonarini et al., 2016	◆		◆				◆				◆	◆					◆					◆		◆	
Bondioli et al., 2019	◆	◆		◆	◆		◆	◆	◆		◆							◆			◆	◆		◆	
Dziobek et al., 2006	◆			◆				◆			◆				◆							◆		◆	
Escobedo et al., 2012	◆		◆							◆	◆						◆			◆				◆	
Feil-Seifer et al., 2009	◆			◆	◆				◆		◆	◆					◆					◆		◆	
Fisicaro et al., 2019	◆		◆				◆				◆						◆					◆		◆	
Garzotto et al., 2017	◆		◆						◆	◆	◆	◆					◆					◆		◆	
Garzotto et al., 2018	◆		◆	◆			◆		◆	◆	◆	◆					◆				◆	◆		◆	
Gelsomini et al., 2018	◆		◆						◆	◆	◆	◆					◆					◆		◆	
Goodwin et al., 2006	◆			◆					◆		◆				◆							◆			◆
Goodwin et al., 2019	◆	◆		◆	◆		◆		◆		◆				◆						◆				◆
Hailpern et al., 2009	◆			◆					◆		◆	◆					◆			◆				◆	
Hayes et al., 2008				◆					◆		◆	◆					◆			◆				◆	
Heathers et al., 2019	◆			◆					◆	◆	◆	◆						◆		◆				◆	
Hirano et al., 2010	◆		◆	◆					◆	◆	◆	◆					◆			◆				◆	
Hong et al., 2012	◆	◆	◆					◆		◆	◆	◆					◆				◆			◆	
Kientz et al., 2005		◆		◆				◆		◆	◆	◆					◆			◆				◆	
Klin et al., 2002					◆						◆							◆		◆					◆
Piper et al., 2006	◆			◆					◆		◆		◆				◆			◆				◆	
Sobel et al., 2016	◆		◆						◆				◆	◆			◆							◆	
Suh et al., 2016		◆		◆		◆	◆	◆											◆					◆	
Tam et al., 2017	◆			◆	◆	◆	◆				◆	◆									◆	◆		◆	
Tartaro et al., 2008	◆			◆	◆							◆				◆						◆	◆	◆	
Wallace et al., 2010	◆			◆					◆			◆			◆						◆	◆		◆	
Whalen et al., 2010	◆	◆		◆							◆	◆			◆							◆		◆	

	Target End User					Setting					Involvement of Person with Autism				Publication Venue					Empirical Support			Tech. Maturity		

Figure 2.1: …continued.

CHAPTER   3

Personal Computers and Multimedia

Use of technologies in support of autism began with the advent of the desktop computer in the late 1970s and early 1980s. The expansion of internet technologies in the last two decades has broadened this scope as well, but we here include those systems with a traditional "desktop" model to them even if people are now accessing them on a wide variety of internet-connected devices. The experience of using these applications usually requires an individual to sit at a screen and use a keyboard and/or mouse to interact with specially designed software in a primarily stationary and seated position. While many of the examples provided in this chapter include applications that have since transitioned to smaller and mobile platforms, we highlight here applications that were initially developed for the stationary platform. A prominent use of the "desktop" platform involves the use of multimedia (e.g., image, video, audio, and combinations thereof) to support teaching and assessment of autistic individuals. This use of video and holds a particularly prominent place in the space of technologies for autism. Although use of multimedia features in other technology reviews in other chapters of this review, in this chapter, we particularly focus on the capture, storage, and/or access of a combination of text, audio, still images, animation, video, or interactivity content forms. It also includes interactive videos, DVDs, or other multimedia, which had an explosion of interest in the late 1990s. We particularly highlight ways in which video is a mode for both collection and delivery of information. We note that due to the changing nature of technology, this chapter is more of a historical overview and less of a review of current trends, as very little recent research in this space has focused exclusively on the personal computer, the web, or multimedia as a platform.

3.1   OVERVIEW OF PERSONAL COMPUTERS, WEB, AND MULTIMEDIA

The first personal computers were made available to the general public in the late 1970s and early 1980s, with platforms such as the Apple II, TRS-80, and Commodore 64. The internet and the World Wide Web (or the Web) became popular a little over a decade later with the invention of the first web browser, Mozilla, in 1992 and the launch of Netscape in 1994. Coincidentally, the rising popularity of computers and the internet followed a similar timeline with that of the rise of awareness and diagnostic rates of autism. Thus, it is somehow fitting that computers have become so popular with individuals with autism, though we are certainly not suggesting this is causal.

For the purposes of this review, personal computers and the Web as a platform includes applications that use a traditional keyboard, mouse, and monitor, and internet-based applications that are primarily designed for access via a computer-based Web browser, and most often accessed today through netbooks or Chromebooks. They can also include laptop-based technologies, but the primary differentiator is those that are intended to be used while stationary and not while mobile.

Applications originally designed for traditional personal computers and the Web initially had particular advantages for their use by autistic people and those who identify as on the autism spectrum or having related experiences. For one, the popularity and prevalence of desktop computers and laptops, in combination with Web technologies, made them one of the first ways to distribute interactive technological interventions to large-scale audiences. Their relative affordability by the 1990s also allowed them to reach the largest audiences of any of the interventions at the time, though the emergence of smaller, mobile platforms by the late 2000s surpassed them. Many schools had at least one desktop computer per classroom, and efforts like One Laptop Per Child[6] have, more or less successfully, increased the penetration of computing in developed nations (see The Charisma Machine for an in-depth look at this particular effort (Ames 2019)). Thus, traditional computers were appropriate for many educational settings where other systems—such as mobile phones—were not yet allowed, affordable, or practical (Cramer and Hayes, 2010). Because computers, particularly desktops, tend to be in fixed locations, they are less likely to be lost, stolen, or broken. Using these systems, people also tend to have a stable place and environment where they can do their work. The systems have also been around for a significant amount of time and thus have a long history of devices and software that can be run on different platforms, and often comprise more powerful systems that can run more data-intensive and processing-heavy programs without having to worry about battery life or network connectivity. Use of the Web has a unique advantage of being able to be accessed from almost any computer, including many mobile devices. They can also allow for instant updates without requiring users to install upgrades. Finally, for the most part, Web applications do not depend on any minimum system requirements.

On the other hand, desktop-based systems requiring users to be in a fixed location necessarily limit their flexibility to fit into everyday activities. Even laptops, which are more mobile than desktops, typically require a certain amount of space within which to operate and can include cumbersome peripheral devices. In addition, many applications built for traditional computers require the use of a mouse, keyboard, or both, which may have a steeper learning curve for people with autism if they are non-verbal, have motor impairments, or experience other barriers to the traditional computing experience. Personal computers and the desktop paradigm also may not be able to take advantage of some of the more interactive, real-world technologies described in other sections, which may make generalization of skills learned on a computer more difficult than it might be with other technologies.

Another focus of the review of this chapter is on the use of video and multimedia. Given that autism is largely characterized, diagnosed, and monitored behaviorally, and that autism tends to be

---

[6] http://one.laptop.org/

associated with visual learning (Quill, 1997), multimedia—particularly images and video—is an appealing mechanism for collecting, analyzing, and delivering vast amounts of information. Baskett (1996) showed that having video conversations with another individual reduced anxiety, and thus video technology can serve as a useful intermediary. This may have implications for people who use more modern video chat technologies, such as Skype or Google Hangouts, as Mokashi et al. (2013) explored with children. This approach is not without its challenges, of course, including technical as well as social, practical, and legal concerns. However, even with these challenges, the affordances, opportunities, and potential for video in support of individuals with autism is so strong that a large number of research projects have focused on its use. Before we describe those projects in depth, we first begin with some basic definitions and background on video.

Most dictionaries define video as relating to the recording or playing of "moving visual images", with multimedia expanding this use of video to include references to computer text. We focus on using these relatively broad considerations of video and multimedia as a platform for diagnosing, monitoring, instructing, and supporting individuals with autism. However, such a broad definition necessarily requires that we not include every possible research project or commercial product in existence.

Much work has been done to capitalize on the visual processing strengths observed in individuals with autism in terms of instruction and reduction of behavior deemed problematic by the dominant culture. At the same time, with no specific genotype or physiological phenotype for autism, researchers and clinicians rely on behavioral phenotyping for diagnosis and monitoring. It has been argued that video and multimedia records can become essential parts of this process through what some reference as "behavior imaging" (Narayanan and Georgiou, 2013), an analogy to medical imaging that refers to the capture and recording of behavioral data. The concept of multimedia is changing, as elements like games and immersive environments become more plentiful. For those platforms, however, we direct readers to Chapters 6 and 8 in this book.

Across this broad space across personal computers, the Web, and the use of multimedia and video, we group technologies in this chapter based on the following categories: (1) specialized software or websites designed specifically for autistic individuals; (2) mainstream software, websites, or other technology applications specifically used by individuals with autism; (3) studies comparing computer-based tasks to other types of interactions; (4) multimedia instructional aids; (5) interactive multimedia; and (6) use of video tools for recording and assessment.

3.2   PERSONAL COMPUTER AND MULTIMEDIA USE AND AUTISM

3.2.1   SPECIALIZED SOFTWARE AND WEBSITES

A number of researchers have designed, built, and studied the use of specially designed software for individuals with autism and their caregivers. These applications address a wide variety of areas from education to screening to communication.

One of the earliest examples of specialized educational software for autism is DT Trainer[7] by Accelerations Educational Software. DT Trainer is a computer-aided instruction program that uses components of Applied Behavior Analysis (ABA), a popular intervention for individuals on the autism spectrum (Lovaas, 1987). This was one of the first commercially available and low-cost systems to support discrete trial training in school systems and has been deployed at a relatively low cost across many schools. As ABA receives more scrutiny (e.g., Goldiamond, 1974; Milton, 2018; Shyman, 2016) and evolves (Leaf et al., 2016), we expect DT Trainer to evolve as well. A more research-based application in this space is TeachTown (Whalen et al., 2006), based on a variant of discrete trial training, called Pivotal Response Training (Koegel et al., 1999). Teach-Town has been subject to rigorous validation for efficacy (Whalen et al., 2010). Researchers using TeachTown conducted a between-subjects study across 47 classrooms in the Los Angeles School District and determined that those who used the software across the three-month period showed more improvement on cognitive and language outcomes than those in the control group. The basics of this approach have now been documented for use by others as well as some reflection on the changing nature of these behavioral approaches and changing views of ABA (Jones, Wilcox, and Simon, 2016).

KidTalk (Cheng et al., 2002) is another program that provides a means of conducting online therapy for individuals with high-functioning autism and Asperger's Syndrome. KidTalk consists of scripts for interaction and provides different rewards for progress and engaging in socially appropriate behavior. It also provides therapists with tools for group therapy and feedback. KidTalk has been tested for feasibility with a number of small groups and shows promising results, but it has not been released as a commercial product. Other serious games like KidTalk expanded from this work (e.g., Grossard et al., 2017).

Teaching facial processing and emotion-recognition skills to individuals with autism or who have other sensory processing challenges using software has been a common thread of work in this space. Tanaka and colleagues (2003) have developed a framework for studying facial processing defects in individuals with autism they call Let's Face It! This system is made up of a series of mini games designed to teach facial processing, including finding a face in a scene, matching faces with

---

[7] http://www.dttrainer.com

similar expressions, creating a face with different emotional components, and following eye gaze of a face on the screen (see Figure 3.1, left). Bölte and colleagues (2002) designed and evaluated a system (see Figure 3.1, right) with 10 individuals with autism to teach recognition of facial affect, based on concepts related to Theory of Mind (Baron-Cohen et al., 1985). They found the system produced improvement in detection of facial affect, but they also cautioned against generalization of their findings.

Figure 3.1: Left: "The Eyes Have It" mini game as part of Let's Face It! (Tanaka et al., 2003) and Right: a tool for teaching facial affect recognition (Bölte et al., 2002).

The Mind Reading software application (Baron-Cohen et al., 2004) (see Figure 3.2) was designed to teach people to read and understand emotions by using strength areas typical of autism to address reading emotions. Several studies have identified that individuals with autism have improved emotional response scores after using this software after a set period of time (Golan and Baron-Cohen, 2006; LaCava et al., 2007, 2010). FaceSay, another software application that uses realistic avatar assistants in an interactive, structured environment (Hopkins et al., 2011), was used to teach social skills to children with a range of disabilities and symptoms, all with an autism diagnosis. Their randomized controlled trial study showed that "higher functioning" children improved in facial recognition, emotion recognition, and social interactions while "lower functioning" children improved in emotion recognition and social interactions, making it a promising application. A similar research prototype, called Facial Expression Wonderland, has similar goals for teaching facial expressions to children with autism (Tseng and Do, 2010). The system has not yet been evaluated rigorously at the time of this writing; however, Ould Mohamed and colleagues (2006) developed a method for assessing attention analysis in the use of software by combining gaze direction and face orientation. This type of analysis could be used to evaluate all the systems mentioned above.

Figure 3.2: **Mind Reading** (Baron-Cohen et al., 2004).

Other desktop and Web-based software has been designed to improve the ability for people with autism to communicate with people from the dominant communication paradigms and cultures in which they live. For example, Hetzroni and Tannous (2004) developed an application based on daily life activities like play, food, and hygiene to teach specific functions of communication. Specific outcome measures included factors related to echolalia, relevant and irrelevant speech, and communication intentions—all of which showed better alignment with dominant cultural norms after using their system. The authors claim that their software also generalized learning to the traditional classroom environment. Lucas da Silva and colleagues (2011) developed a software application for improving communication skills called TROCAS (see Figure 3.3). which supports photos, audio, videos, a message board, an online digital book library, and connection to a story tool. The design of the system was specifically built to allow autistic people some level of agency by specifying communication preferences and presenting them in a format that would allow for the best user experience. Preliminary tests of the use of TROCAS by 3 children across 12 weeks showed some improvement in communication competencies.

Figure 3.3: **TROCAS system** (Lucas da Silva et al., 2011).

In addition to software designed specifically for use by individuals with autism, websites and software have been designed for caregivers of people affected by autism in other ways. For example, the Autism Support Network[8] and now-retired TalkAutism[9] have been places for parents and caregivers to find resources, obtain peer support, ask advice of experts, and seek advocates. Baby Steps (Kientz, 2012; Kientz et al., 2009) was a research platform designed for parents of young children to help track developmental progress and help detect autism (among other developmental delays) earlier (see Figure 3.4). The system was a stand-alone application that asked parents to track milestone questions along with sentimental memories of a child's younger years. Baby Steps was evaluated with eight families for a three-month period of time and findings suggested that certain features allowed for timelier tracking of milestones and improved communication with pediatricians. It later evolved to a website, text messaging platform, and integration with social media sites like Twitter (Suh et al., 2014, 2016). The text messaging platform for Baby Steps is further described in Chapter 4.

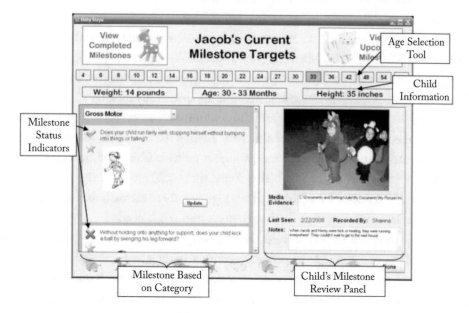

Figure 3.4: Baby Steps system for tracking developmental milestones in young children as a software platform (Kientz et al., 2009) and then evolved as a website (Suh et al., 2014) and eventually into a text messaging platform (Suh et al., 2016).

[8] http://www.autismsupportnetwork.com/
[9] http://talkautism.com

3.2.2   MAINSTREAM SOFTWARE AND WEBSITES

In addition to the numerous systems specially designed for use by autistic people, there have also been uses and adoption of more general-purpose software or websites by those experiencing or identifying with autism. These include tools for computer-mediated communication as well as educational or workplace technology.

People with autism use a variety of tools for communicating. For example, Burke and colleagues (2010) conducted a qualitative study on how computer-mediated communication (CMC) tools are used for social purposes by individuals on the autism spectrum. Their study examined uses of cell phones, text messaging, email, instant messaging, social networking sites, online dating sites, and discussion forums. Their analysis revealed that many individuals with autism, despite social connectedness being an area of deficiency as part of their diagnosis, are able to use CMC tools to connect with others and develop successful relationships. However, their findings indicate that people with autism may still have issues relating to trust, disclosure, inflexible thinking, and perspective taking, making it harder for them to maintain relationships.

As social networking sites have increased, so has research interest in studying its impacts on individuals with autism. In 2013, Mazuerk (2013) specifically explored autistic adult use of social networking sites and found positive correlations with friendship quality and quantity. Mazurek and colleagues (Ward et al., 2018) later followed this up, specifically looking at how Facebook and Twitter impacted happiness, and found that adults who used Facebook were happier than those who did not, but there was no such relationship with Twitter. Another analysis explored specifically explored adolescents' use of social media and its impact on friendships and how it was affected by anxiety levels (van Schalkwyk et al., 2017). They found that social media use by autistic adolescents without anxiety could enhance friendships. Taken together, there seems to be a positive impact of certain types of social media experiences, though more research is needed.

Benford and Standen (2011) studied the feasibility of using email to facilitate interviews with 23 autistic adults, as opposed to in-person interviews. In their paper, they discuss epistemological, methodological, and practical issues of doing research in this way, and found that it can provide a viable means of conducting research with a population that is often difficult to reach, and one that allows them to have greater control over what they say and how they say it.

Educational games and software designed for use by students in general have also been used in a number of studies with individuals with autism. For example, Lewis and colleagues (2005) used The Learning Company's Clue Finders game as a tool for assessing how children worked on computers in social interactions and influence acceptance by peers in classroom interactions. Pixwriter allows writing along with pictures in a split-screen format to teach students with autism to write using a computer (Pennington et al., 2012). Early learning reading software called Headsprout helps teach word lists and text reading skills (Whitcomb et al., 2011). Our experience in working with educators working with autistic students suggests that there are many uses of mainstream software applications beyond these specific examples and that, in general, educators should be encouraged to be flexible and creative in developing accommodations and using instructional aids. However, one challenge is that there are often difficulties in finding age- and skill-appropriate software for a given classroom and for a given individual.

Beyond tools specifically developed for education and then repurposed for use by people with autism, there have also been investigations of technology tools developed for general use. Several researchers have used an older tool called Bubble Dialogue (Cunningham et al., 1992), developed in 1992 as a means of using a HyperCard-based technique as a constructivist education tool that uses a comic-strip model with speech bubbles to teach communication skills (see Figure 3.5). This tool has been used successfully by autistic people across several different projects. For example, Dillon and Underwood (2012) determined children with autism could use Bubble Dialogue in their process of imaginative storytelling. Glenwright and Agbayewa (2012) used the tool to help higher-functioning children and adolescents to comprehend verbal irony through computer-mediated communication. Similarly, Rajendran and colleagues (2005) used the tool to explore how individuals with Asperger's Syndrome can use CMC tools to respond to non-literal language and requests made through technology that others may deem to be inappropriate. Although Bubble Dialogue is nearly 30 years old, recent studies still find that it is a useful tool for understanding computer-based interaction skills that may generalize to other types of CMC applications.

Figure 3.5: Bubble Dialogue application (Cunningham et al., 1992).

Google SketchUp is a 3D modeling tool that researchers have identified as a useful and attractive way for engaging the visual-spatial skills of individuals with autism. Wright and colleagues (2011) conducted a qualitative study of Google SketchUp in seven boys with high-functioning autism and found that it promoted intergenerational communication between parents and grandparents. And although video games are across many platforms, they do include PC-based games, and thus Mazurek et al.'s (2015) study of autistic adult experiences with mainstream games is relevant here. They found that adults are motivated to use video games to reduce stress, be immersed, to pass the time, and for social connections, but concerns about addiction, negative social interactions, violence, sexual content, and poor design were surfaced.

Beyond individuals in Burke and colleagues' study mentioned previously (Burke et al., 2010), there have been some popular examples of individuals with autism using computer-mediated communication. For example, there is a popular website, WrongPlanet.net, which serves as a blog, discussion forum, and chatroom for individuals with autism and their families, and the popular virtual world, Second Life, has a number of worlds dedicated for people with autism, and Stendal and Balandin (2015) specifically focused on Second Life in a case study with an autistic individual who found communicating through the platform to be more comfortable. There have also been uses of blogs and YouTube videos of individuals with autism. One YouTube video, "In My Language" by Amanda Baggs, was used to explain how she experiences the world, and has been viewed over 1 million times.[10] There are likely many more uses of mainstream software and websites beyond those mentioned, but in our literature search, we did not identify many research studies exploring them. The number of websites, blogs, social networks, and mainstream software applications are voluminous. There is certainly room for additional research to explore which tools are the most useful, why, and how they can be improved and made even more useful for individuals with autism.

3.2.3   COMPARISON OF COMPUTER-BASED TASKS WITH OTHER TYPES OF INTERACTIONS

Perhaps some of the earliest uses of computer-based applications for individuals with autism are those involving simple computer-based substitution for tasks typically accomplished without the use of digital technologies. The reason for these types of comparisons is the conjecture that including humans or physical components in educational tasks can provide too much stimulation for many autistic people to concentrate, and thus they may have better performance on a computer. There is also motivation to move certain interventions to the computer, as it can be expensive for one-on-one individualized interventions with a human being, and thus a child with autism could receive more therapy in a given week if some of it was conducted via computer.

There are mixed results in studies that compare the use of technology compared with in-person interaction. For example, nearly 35 years ago, Plienis and Romanczyk (1985) conducted a study that looked at the performance and behavior of individuals with learning and behavioral problems (including autism) when they engage in a two-choice discrimination task. They found no

---

[10]   http://www.youtube.com/watch?v=JnylM1hI2jc

differences in task performance between the computer and adult delivering instruction, but they did observe more behavioral difficulties when they engaged in the task with an adult. About 10 years later, Chen and Bernard-Optiz (1993) compared computer-aided instruction with teacher-based instruction. They found computers to be more interesting and motivating, eliciting better behavior during computer use. Another later study by Moore and Calvert (2000) specifically compared a computer-based application and a behavioral program with a teacher for teaching vocabulary tasks. They found the autistic children to be more attentive, motivated, and able to learn more vocabulary with the computer program. Baker and Krout (2009) also explored how video communication technologies like Skype can be used as an intervention using music therapy. Taken together, it appears there is promise in computerizing tasks traditionally performed by a teacher or therapist for children with autism, promise that should be explored further in and adult learning context.

Another study tested three different types of instruction for teaching non-verbal reading skills to three different students, one of whom had autism (Coleman-Martin et al., 2005). The three cases included teacher-only instruction, teacher plus computer, and computer-only instruction (using Microsoft PowerPoint™) and found that all three students met criteria equally well across conditions. An additional look at reading skills (Williams et al., 2002) compared an interactive and engaging book-based experience with a scanned and digitally enhanced (with sound files, interactive elements, etc.) through a multimedia authoring tool called Illuminatis. After 10 weeks, they found that children with autism spent more time on the computer task, and 5 out of the 8 children tested acquired new words. More recently, Barrow and Hannah (2012) conducted an exploratory study looking at computer-assisted interviewing as a means of gathering input and feedback from individuals with autism. They reflect on their experiences compared to traditional interviewing techniques and conclude that it is a viable tool for communication.

Researchers have also explored how autistic children compare to typically developing peers during computer-based tasks. For example, Bernard-Opitz and colleagues (2001) compared the use of a computer program to teach social problem solving by eight matched pairs of individuals with autism and typically developing peers. They found that children in both groups could be taught problem-solving skills through software, although the individuals with autism learned at a slower rate and produced fewer results.

Other research has compared the use of computer instruction for not just individuals with autism, but also their therapists. For example, Granpeesheh and colleagues (2010) studied the use of an eLearning tool for conducting training for applied behavior analysis therapy compared to traditional one-on-one, in-person instruction. They found similar results in knowledge acquisition for both methods, but that in-person instruction resulted in slightly better scores. However, they conclude that eLearning strategies can be useful when resources are not available for in-person instruction. Another study compared inexperienced teachers conducting therapy when given a computer-assisted instructional task versus one-on-one instruction and found that teachers had

90–100% accuracy in the computer-based task compared to only 60% in the one-on-one instructional task (Kodak et al., 2011). This demonstrates that there is promise in using computers not only for individuals with autism, but also their therapists and teachers.

Other research has shown that computer-based activity schedules are successful in teaching individuals with autism a variety of new skills through the use of videos, sounds, and images (Stromer et al., 2006). These activity schedules have been created using standard prompts as well as Microsoft PowerPoint (Rehfeldt et al., 2004). More recent research (Constantin et al., 2017) has explored how computer-based rewards are best received by autistic children (along with children with intellectual disability) and provide recommendations for computer-based reward systems.

3.2.4   MULTIMEDIA INSTRUCTIONAL AIDS

In a variety of classroom environments, video has long been used to teach an assortment of concepts, whether through watching a science demonstration or learning appropriate social skills. We describe video-based instruction separately from the broader category of multimedia, which may include video but does not have to, and may include, other media elements. Finally, we close with a discussion of tools to support authoring of these instructional media, a continued challenge for those wishing to use them.

Video Modeling and Image-Based Instruction

Some learning theories argue that observing behaviors and attempting to replicate or model them account for much of the traditional pathways to acquiring specific skills (Shipley-Benamou et al., 2002). For autistic learners, application of these approaches more directly, through video modeling and video-based instruction, can assist in learning and retaining positive behaviors (See Figure 3.6 for an example). Several articles have reviewed in-depth the area of video modeling for autism (Ayres and Langone, 2005a; McCoy and Hermansen, 2007; Bellini and Akullian, 2007; Delano, 2007), and the summary of salient points follows.

Video modeling, a type of video-based training, typically involves watching recorded videos of others or oneself performing a behavior correctly or positively. The former, peer modeling, focuses on observing people similar to oneself (in physical characteristics, age, group, ethnicity, etc.), whereas the latter, self-modeling, encourages watching oneself complete a task successfully. It can also be helpful to view the world as though one is experiencing it, a so-called "first-person shooter" perspective (Shipley-Benamou et al., 2002).

Video modeling traditionally focuses on only watching those videos in which the person performs the activity correctly. However, recording and watching videos of oneself performing the same activity over time opens up opportunities for self-evaluation and visual progression of personal growth over time. Likewise, using video to learn and practice skills enables delivery of positive feedback and presentation of concepts and instruction in engaging ways (Goldsmith and LeBlanc, 2004).

Finally, the use of video training, both video modeling and other kinds of instruction, appears to be effective, at least in part, because it directs focus to relevant stimuli by requiring the viewer to look at a small spatial area and listen to accompanying language (Shipley-Benamou et al., 2002).

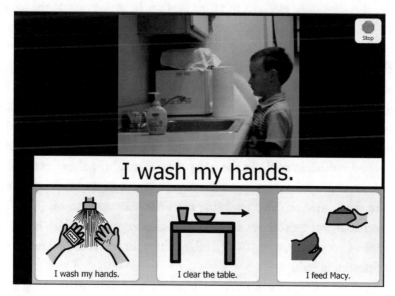

Figure 3.6: Video-based modeling feature from Boardmaker Studio software.[11]

The availability of relatively low-cost mechanisms for recording and replaying videos has enabled research to be done on the effectiveness of video-based instruction for individuals with autism for more than two decades. Ayres and Langone (2005a) reviewed much of this literature, concluding that more detailed studies would be required to describe the specific mechanisms by which video-based instruction works, but generally speaking find that parents, educators, and other caregivers can effectively use video for instruction. Video-based instruction has been shown to successfully teach a variety of skills, including developmental, life and transition, as well as speech and language and even academic skills. In what follows, we briefly overview some of the work in these three areas demonstrating the ways in which video-based instruction can and should be used with individuals with autism, as well as open questions still to be explored.

Video modeling appears to improve in-vivo instruction when teaching developmental skills and enhancing generalization of those skills outside the academic instructional environment. Additionally, classroom management issues, such as reducing disruptive behavior during transition time, have been shown to benefit from video priming (Schreibman, et al., 2000). However, evidence for

---

[11]   https://goboardmaker.com/products/boardmaker-studio

this improvement is still somewhat scarce and needs replication. One study, comparing video modeling with in-vivo modeling, focused on five students with autism who all experienced a baseline condition of prompted responses, rewards, and other best practices followed by either video modeling or in-vivo modeling of a specific task (Charlop-Christy, et al., 2000). In this study, students all performed their tasks independently after the same or fewer minutes with video modeling than in-vivo and showed at least some evidence of generalization. In another study, the focus was only on one preschooler with autism, but a multiple baseline design enabled the researchers to suggest that video modeling supported generalization of skills mastered in the academic environment only (in this case, preschool songs and other age-appropriate instructional aids) (Kleeberger and Mirenda, 2010). One challenge to the study of developmental skills in general, but in particular in response to a technological intervention like video modeling, is the choice of tasks and the preparation of materials—videos—to support those tasks. These tasks should be similarly difficult across participants but also of specific need for the child being tested, which makes it virtually impossible to establish a fully controlled and consistent study. In the two papers highlighted above, something closer to a case study model was employed for analysis even though in one of the studies the tasks were randomly assigned to the in-vivo or video modeling conditions for comparison (Charlop-Christy, et al., 2000). New technological approaches, such as crowdsourcing videos, which has been done when writing social stories (Boujarwah, et al., 2012), might allow for much larger studies with a greater breadth of participants and tasks.

Tightly related to concerns about teaching developmental skills are teaching life skills, which are key to transitioning to independent living. Researchers and clinicians alike have long used static picture-based prompting to teach a variety of life skills in autism, but a direct comparison of picture and video-based prompting indicates that video prompting may be more effective, as well as less expensive, to implement and deliver than picture-based approaches (van Laarhoven et al., 2010). Various approaches to video-based instruction for life skills have been tried and shown to be successful, including standard observational or training videos as well as more specific video-modeling approaches. These tools are rarely used in isolation, however. For example, a video instructional package for teaching grocery shopping skills was found to be effective for three children with autism but included on-site prompting and reinforcement in addition to video training (Alcantara, 1994). Likewise, an image-based tool for teaching photography skills to adults with developmental disabilities showed that video prompting could be an effective instructional strategy, allowing for both generalization and skill maintenance over time (Edrisinha et al., 2011). The advent of new mobile technologies that are commonly used and less stigmatizing than special-purpose assistive technologies has enabled the use of video modeling in general education (Cihak et al., 2009) and community-based environments (Nikopoulos et al., 2008). Similar to teaching developmental skills, however, a major limitation to teaching life skills through video—and subsequently conducting research about its efficacy—is the creation of large libraries of instructional videos. A variety of computational approaches, including programmatically changing backgrounds, actors (or at least skin and hair color), and other elements of videos may enable larger corpuses of realistic life skills videos to be produced that can facilitate larger trials to assess the efficacy of these approaches. Although limited in scope, some research has attempted to demonstrate video modeling as a tool for reduction of certain behaviors, rather than just an increase in other behaviors. In one study of three children with autism using a multiple-baseline-across-subjects design, video modeling was shown to be effective at reducing problem behaviors for children with lower baseline levels of disruptive behavior (Nikopoulos et al., 2008).

Given the prominence of instructional concerns around speech, language, conversation, and social skills for individuals with autism in general, it is unsurprising that video-based instruction has evaluated these issues as well. In fact, a review published a few years ago provides a nice overview of this literature (Shukla-Mehta et al., 2009). A variety of programs have been developed to enhance speech and language skills in individuals with autism, many of them taking advantage of the affordances of video and multimedia. For example, in a study of 20 children with autism and mixed intellectual disabilities and 9 teachers using a specially developed multimedia program, an overall increase in verbal expression was found. Those with low language also showed an increase in verbal expressiveness, while those with high language showed an increase in enjoyment (Sherer et al., 2001), indicating that while the effects may be slightly different between groups, such programs can be used across the spectrum of verbal capabilities in individuals with autism.

Although video modeling and video-based instruction have not been used extensively to teach academic skills to children with autism, the limited available research is promising. Fourth-through sixth-grade students with learning disabilities showed statistically higher word-acquisition scores when exposed to a video instruction program than their peers who received no video instruction (Xin and Reith, 2001). In another, more limited, study involving only one child with autism, the student watched a variety of video models, including a teacher writing the word and play acting a word's meaning, eventually learning to spell enough novel words to match her general school placement (Kinney, et al., 2003).

Video and multimedia have also been shown to support teaching social-language skills to individuals with autism (Maione and Mirenda, 2006). In some cases (e.g., Sansosti and Powell-Smith, 2008), traditional tools like social stories (Gray, 2003) are augmented by video and other media. In other cases, video modeling, in its more traditional form of scripted videos to be watched before an interaction, is shown to be effective in teaching social skills. For example, one study of two children with autism who watched videos prior to interacting demonstrated that they made more appropriate play comments when engaging in play sessions with their siblings (Taylor et al., 1999). As another example, the VidCoach system supports learning an important transition-related social skill, job interviewing (Ulgado, 2013). Use of this system over multiple weeks on an ad hoc basis resulted in improved outcomes in terms of interviewer assessment as well as research coding of interview performance in job interviews with actual employers (Hayes et al., 2015).

While promising, the effects of multimedia training are not dramatically better than those observed in conventional therapist-instructed training. Additionally, a therapist can respond in situ to the particular needs and proclivities of each student, whereas customization might be required for the video-based tools to provide this same level of support (Wong and Tam, 2001). This customization may include changing elements in the video to match the context of activity for that student, such as outdoor classrooms in California and Florida as opposed to enclosed school hallways in other areas, or the context of the students themselves, such as ethnicity, gender, or even height of the students. These customizations require substantial content generation and can be challenging. Other customizations, however, can be accomplished technologically. For example, one study showed that although students with autism performed worse than their neurotypical peers on video-based emotional and facial recognition, their performance improved in relationship to the speed (slowness in this case) of the video (Tardif et al., 2007).

3.2.5   INTERACTIVE MULTIMEDIA

Both researchers and practitioners have begun to expand traditional visual supports using multimedia (Hayes et al., 2010) (see Figure 3.7), often coincidentally and opportunistically (Stromer et al., 2006). For example, Stromer and colleagues' (2006) review notes that the expansion of activity schedules (an exemplary visual support) through computing technologies enables learners to develop new skills through audio, video, images, and coordinated text. These effects were present on sometimes elaborate but often fairly simple computerized visual supports, leading to the conclusion that interactive multimedia simultaneously presents opportunities for teaching generative and functional skills and provides a framework for understanding acquisition of these skills.

Figure 3.7: Interactive visual supports (Hayes et al., 2010).

Just as with video-based instruction, interactive multimedia has been used to support a variety of therapeutic and instructional interventions, including play (Dauphin et al., 2004) and other social skills (Hagiwara and Myles, 1999; Kimball et al., 2004), activities of daily living (Rangel and

Tentori, 2011), and academic skills like reading (Heimann et al., 1995) for autistic children. Although many of these projects primarily include a mix of video and activity schedules (e.g., Kimball et al., 2004), other modalities are also used. For example, photographs as part of a computer-based instructional module were shown to be helpful in teaching 11 children with autism or Asperger's Syndrome to recognize and predict emotions after using the program for 5 hours spread across 10 sessions in 2 weeks (Silver and Oakes, 2001). As another mixed-media example, the Mind Reading DVD uses silent films of faces, video recording, and written examples of situations that evoke particular emotions to teach emotions and mental states (Golan and Baron-Cohen, 2006). As one final example, text, speech, and images were combined to make a set of games that target specific communication disorders. In a study of 10 adolescents with high-functioning autism and 10 neurotypical adolescents, the researchers found that richer multimedia interfaces were more challenging for students with autism, indicating that perhaps there is a limit after which additional multimedia becomes more of a hindrance than a help (Grynszpan et al., 2008).

Multimedia Authoring Tools

Many multimedia approaches rely on teachers or other caregivers and educators to become adept not only at creating engaging and informative lessons, but in doing so through technology. Fear, or even a simple lack of technical skills, can prevent this kind of innovation. However, only limited work has investigated how to author these materials more easily.

Higgins and Boone (1996) present a set of software guidelines based on their research on two multimedia authoring systems: HyperStudio and Digital Chisel. Although both of these systems are now nearly 20 years old, the concepts of how to invoke certain lessons within these paradigms may still be useful to both educators and researchers seeking to create such materials. Their very pragmatic instructions, such as buying as many computers as can be afforded, likely hold true today in most settings. Likewise, their suggestion of drawing out the lesson as it would appear in the software program is a useful technique for both software developers and lesson planners.

More recently, advanced computational techniques have been explored to make authoring these materials simpler and more efficient. For example, both artificial intelligence (Riedl et al., 2009) and crowdsourcing (Boujarwah, 2012; Boujarwah et al., 2012) (Figure 3.8) have been demonstrated in limited trials to enable the production of large quantities of high-quality social stories. Some researchers have explored how to enable students with autism themselves to create and use their own multimedia skills in developing social skills training materials (Cumming, 2010). Future work is required to test materials derived from these approaches in use, but the promise of—and need for—large quantities of freely available teaching materials for social skills makes it likely that the work will continue.

Figure 3.8: Crowdsourcing-generated instructional models (Boujarwah et al., 2012).

3.2.6   RECORDING AND ASSESSMENT

Although the bulk of work on video and multimedia for autism has focused on using multimedia tools as output, or instructional materials, researchers and providers alike have become increasingly interested in using them as input to clinical and care processes. Video can be used as standardized assessment tools, in addition to questionnaires and behavioral measures already in place. It can also be used to capture and document activities in support of diagnostic or monitoring efforts. In this section, we describe some of the research efforts focused on using video for assessment and record keeping.

Assessment of Interactions via Video

Just as showing videos can be useful for instruction, asking questions about what was discerned from video can be useful for assessment. For example, the Movie for the Assessment of Social Cognition (MASC) toolkit includes a short video that participants watch followed by a set of questions to be answered that reference actors' mental states (Dziobek et al., 2006). In a study involving 39 participants, researchers found MASC to discriminate individuals with autism from controls. Likewise, Golan and Baron-Cohen (2006) demonstrated with a similar sample size that adults with autism performed significantly lower than controls on questions describing mental states of actors in social scenes from feature films.

A wide variety of tools already exist for video annotation, such as Elan[12] and Anvil.[13] Additionally, researchers have explored specific issues related to video coding as part of assessment and scientific work. A common trend in technology research, particularly early stage, is to video record the use of technologies in a laboratory setting. Hailpern et al. developed the A-cubed method for

---

[12] http://tla.mpi.nl/tools/tla-tools/elan/
[13] http://www.anvil-software.org/

coding such videos in an attempt to provide some standardization across reporting of their usage. They include audio and vocalization elements, physical interactions, as well as specific details about responding to interactive systems. Their coding scheme provides high reliability in their experiments but requires intensive engagement with videos (between 20–40 minutes of work for each coder per minute of video) (Hailpern et al., 2008).

VideoCapture

Manually recording data, whether using pen and paper or digital tools, can distract recordkeepers from their primary activity of interest. Thus, to support the need for extensive documentation of a wide variety of behavioral phenomena, researchers have investigated the application of capture and access technologies (Truong et al., 2001) to the recording of educational, behavioral, and health data about and for individuals with autism (Hayes et al., 2004; Kientz et al., 2007). Truong and Hayes note four core benefits of using automatic capture technologies, including video recording.

1. Large quantities of rich data can be captured without the need for distracting human intervention.

2. These data can be automatically categorized and tagged, allowing for easier retrieval in the future.

3. People do not have to predict which data will be useful prior to an event, as in manual recording, and instead can determine importance after the event.

4. Automatic tracking of the provenance of data alongside collection of more data from different perspectives can reduce errors and selectivity in records.

Given these benefits and the greater emphasis on use of technology in classrooms and homes in support of autism more broadly, it is perhaps unsurprising that researchers have recently dedicated significant attention to the creation and evaluation of capture and access technologies in this space. In fact, use of these technologies has spawned a new approach to autism diagnosis and monitoring, called behavior imaging (Naranyan and Georgiou, 2013), which involves a collection of tools and techniques that allow researchers, educators, clinicians, families, and individuals with autism to understand and act upon observable human behaviors.

Hayes and colleagues (2004) describe a qualitative field study exploring the use of three early-stage prototype systems for capturing video about children with autism: The Walden Monitor, CareLog, and Abaris. In this work, they describe social, practical, and technological considerations that are key for the design of capture applications in this space.

1. People must be able to record and analyze data iteratively as part of a diagnosis, intervention, and monitoring cycle.

2. Rich data generated by video are particularly important for disorders in which there are limited physiological indicators of progress but numerous behavioral indicators.

3. The task of keeping records must be able to fade into the background through the use of these technologies, allowing caregivers and individuals with autism to concentrate on their primary tasks.

4. Designers must ensure that appropriate controls are in place to establish security and privacy of video data, which is by default identifiable.

5. The financial resources required to deploy and use capture and access technologies must be appropriate given the benefits they provide.

6. Because people will almost certainly continue to need manually recorded data alongside automatically captured data, capture and access systems should enable easy and usable integration of these data.

7. The system architecture should provide modular and distributed capabilities for video recording and other capture devices, storage of data, and interfaces for accessing the data, an issue that has become even more salient with the growth of cloud computing.

8. The system must support a variety of levels of views into and mining of captured data, an issue that brings to mind many of the challenges of "big data" currently being explored by other researchers.

After extensive redesign of these applications (see Figure 3.9), additional evaluation (Hayes et al., 2008; Kientz, 2012; Kientz et al., 2005, 2007), and the creation of new prototype systems, this same group revisited these issues three years later (Kientz et al., 2007), noting some key design considerations: understanding the domain, making changes unnoticeable, simplifying interfaces, and allowing for customization. However, they also note continued design challenges, including the difficulty incorporating children with disabilities into the design process, limitations of currently available sensing technologies, and ethical and privacy considerations inherent to capturing large quantities of high-quality data. As new privacy legislation (e.g., European General Data Protection Regulation and California Consumer Privacy Act) are enacted and engaged, these technologies will have to find ways to appropriately—and legally—keep up.

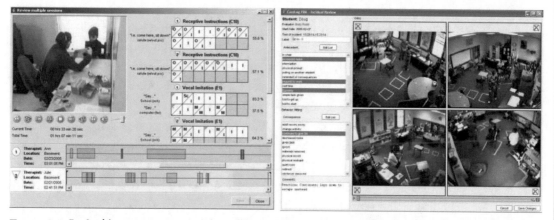

Figure 3.9: Left: Abaris capture tool for applied behavior analysis therapy (Kientz et al., 2005, 2006) and Right: CareLog functional behavioral assessment application (Hayes et al., 2008).

Building on this early work, several research projects have continued to explore the use of video capture for a variety of challenges relevant to individuals with autism. For example, CareLog focuses on the capture of video data for behavioral assessment in classrooms (Hayes et al., 2008) and has inspired a commercial product, BICapture. As another example, Kientz and colleagues developed multiple systems for recording video evidence pertaining to childhood development (Kientz and Abowd, 2009; Kientz et al., 2009). In this work, they describe some of the challenges of recording appropriate "moments of interest" for the complicated task of assessing whether a child may be at risk for a developmental disability. One of the major research questions in this space continues to be how well parents can be trained to collect these research "specimens" at home (Nazneen et al., 2011). Nazneen and colleagues explicitly examined this issue for behavioral problems, finding that parents can be trained in this area, which supports the conclusion made by Kientz and colleagues. Inconclusion, in this section, we have largely described video-capture technologies that are focused on the recording of video data from the environment and direct readers to Chapter 8 on wearable recording and sensing for additional technologies focused on recording video on the body.

3.3   CLASSIFICATION APPLIED PERSONAL COMPUTERS, THE WEB, AND MULTIMEDIA

We here describe how three representative technologies using desktop and Web can be classified in our scheme defined in Chapter 2. This includes TeachTown (Whalen et al., 2010), MASC (Dziobek et al., 2006), and CareLog (Hayes et al., 2008). We also discuss overall trends we observed for technologies making use of personal computers, Web-based platforms, and multimedia.

TeachTown (Whalen et al., 2010): TeachTown consists of a series of educational software applications that are designed for personal computers. The input modality is primarily *physical* due to the use of a mouse and keyboard, and the output modalities are *audio* and *visual*. The domain of skills it teaches are primarily *academic* and *language/communication*. The goal of TeachTown is as an *educational intervention*; however, it is designed for use in both *school* and *home* settings, and thus its target end users are the *person with autism*, their *families/caregivers*, and their *educators*. Teach-Town involved individuals with autism as *users* and *testers* in the development of the application. Whalen et al. published their efficacy study of TeachTown in the journal *Autism*, which is an *autism-specific* publication venue. The TeachTown study in this paper was conducted as a randomized controlled trial, and thus we consider it to be *experimental*. TeachTown is currently *publicly available* via their website.[14]

The MASC Toolkit (Dziobek et al., 2006): The Movie for the Assessment of Social Cognition (MASC) is research tool for assessing "mind-reading" capabilities. This work uses substantial reliance on live-action videos as a key component of the instrument and is intended for use on a personal computer with a mouse and keyboard. Thus, the input modality is *physical* and the output modalities are *audio* and *visual*. This project targets *social and emotional skills* by showing short videos and asking the individual with autism to answer questions about the actors' mental states. The goal of this work was primarily to assess "mindreading abilities in individuals with Asperger syndrome," so this was categorized as *scientific assessment*. The target end users for this work include the *person with autism* themselves, who are tested, as well as a *researcher* who would be collecting the data about the assessment. MASC was developed to be used in a *research lab* and has been used by other researchers, for example to test the effects of medication (e.g., Heinrichs and Domes, 2008) or to compare social cognition with healthy controls (e.g., Pottgen et al., 2013). Individuals with autism have primarily been engaged as users, and the publication venue for this project was in the *Journal of Autism and Developmental Disorders*, which is an *autism-specific* venue. Finally, the studies conducted were *experimental*, and the MASC toolkit at the time of the writing of this paper had reached the maturity of a *functional prototype*, not yet available for public use. Given that other researchers have used MASC, it is now somewhat publicly available. However, the tool is in German and has only been used by German researchers, with English speaking researchers citing it primarily as a motivation for the creation of their own video-based mindreading assessments.

CareLog (Hayes et al., 2008): The CareLog project used video and multimedia as well as personal computers and the web. The use of a physical button by teachers, as well as the mouse and keyboard to use the software, means the input is *physical*. The system relies on multiple audio and video feeds to collect data, and a desktop or laptop computer to store and view both the audio and video data and the meta-data that users associate with them, and thus output modalities are *audio* and *visual*. The assessment function performed by CareLog, Functional Behavioral Assessment, is used to monitor *social/emotional skills*, *restrictive/repetitive behavior*, and *academic skills*. In this paper, teachers were the primary users and gathered data about and assessed progress on all of these skills across student participants. The goal of functional behavior assessment, and thus CareLog, is for *functional assessment* in service of *intervention/education*. One could easily imagine CareLog being used additionally for *parent/clinical training* and other goals; however, in this particular paper, the placement of the study in schools limited the goals that were evaluated. Thus, we did not include other goals in our coding. CareLog was developed for use and evaluated in the *school* setting, with follow-on work expanding it to other settings, and involved children with autism as *users* and *testers*. The work was published originally at the CHI conference, which is a computing publication venue. The research study was a *correlational or quasi-experimental* design, and the maturity of CareLog is a *functional prototype* that is not yet available to the public. However, commercial work building on CareLog is available to the public as described earlier in this chapter.

In general, this chapter includes some of the largest variety in terms of publication venues, because many of the research studies were rigorous evaluations of software, rather than aiming to create a more technically novel system, as many contributions in the computing publication venue strive to do. We saw more publicly available software in this category, as many technologies were easier to distribute via the web and did not require specialized hardware beyond widely available personal computers. The domains and goals for these applications were quite varied, as many software and web applications exist. However, educator end users and school settings seemed very prominent in this category, perhaps due to the widespread availability of traditional personal computers in classrooms and the relatively low cost given that install base. Finally, we saw a higher proportion of experimental studies of technologies in this platform, which may be a result of seeing a higher prevalence of publications in domains outside of computing, where experimental studies are more of the norm.

Multimedia has been used by parents, therapists, community members, and autistic people for decades. Long before computing researchers discovered autism as a salient and important research domain, autism researchers, providers, and educators were making use of advanced (for the time) video and multimedia technologies in the form of films, tapes, and so on. As multimedia has become cheaper and easier to both create and to consume, research projects and practical applications in this space have grown accordingly. At the same time, researchers in computing have long been interested in video and multimedia in terms of production, consumption and analysis, and applications. Thus, the recent intersection of these two areas has resulted in an enormous amount of research in both behavioral/social science venues and those in computing specific conferences and journals.

Given the wide variety of people engaging in this kind of work, it is perhaps not surprising that there is also a wide range of maturity in the technologies represented. Researchers whose primary focus lies in autism, education, psychology, and other related fields tend to use commercially, or at least publicly, available robust technologies and conduct experimental research. Computing specialists, on the other hand, tend to publish work centered on *conceptual* or *functional prototypes*, most of which are not yet publicly available and may never be.

Multimedia, when used for instruction, is primarily targeted at the person with autism or the training of non-professional caregivers. The recording of video and other types of media, however, can be used for a variety of purposes, including but not limited to assessment of professional providers, assessment of the autistic individual, and collection of epidemiological and population-level data.

The domains and goals that can be supported through video and multimedia are substantial and diverse. In our review, there were papers that fit every category of each of these sections in our categorization, and no particular category dominated.

3.4   FUTURE DIRECTIONS

Because of the surge in growth among other technology platforms discussed in future chapters in this volume, growth in the use of desktop technologies has been less pronounced. However, there is still much promise surrounding the potential for desktop, laptop, Web, and social media technologies. Computers are likely to be present in school settings for decades to come because budgets often do not allow for adoption of newer technologies at a fast rate. Likewise, increasing numbers of employers demand extensive computer use across a wide variety of jobs and professions. In addition, websites and social media can often be used on multiple platforms, including shared active surfaces (Chapter 5) and mobile devices (Chapter 4). We anticipate seeing increased use of Web-based applications, as they can run on many devices, do not rely on computers being configured in a certain way, and are easier to update. In addition, some of the limitations of Web-based applications, such as access to specific interactive components, have been addressed as Web browsers and technologies have advanced. The advent of responsive design (Marcotte, 2011) also allows Web designers and developers to make one website that functions on a variety of devices (desktop, tablet, smartphone, etc.) using a single source code—a distinct advantage over previous Web design activities.

More studies are needed to understand the long-term impacts of technology use by autistic people, just as such studies are needed for the general population. Finkenauer and colleagues (2012) conducted a longitudinal study of the use of the internet by those with autistic traits compared to neurotypical individuals. They found that while the number of hours was not different between the two groups, those with autistic traits were more likely to compulsively use the internet. Likewise, a study by Burke and colleagues (2010) found that individuals with autism often had difficulty with trust, disclosure, inflexible thinking, and perspective taking on the internet. However, more recent studies (e.g., Jensen et al., 2019) indicate that much of the concern over youth using digital technologies has been overblown. As with any type of tool, the trade-offs should be considered, and the individuality of the person using it should be taken into account prior to use.

The research conducted thus far around the use of video and multimedia in support of autistic individuals and their support networks covers a wide breadth of areas. Broadly speaking, this body of work indicates that multimedia can be effective both for instruction and for documentation and review. We expect work to continue in this space, both as the research projects mature and as commercial products become more prolific and robust.

Multimedia is becoming increasingly commonplace online, in educational software, and in our everyday lives. For children with and without autism, there is no doubt that there will continue to be a proliferation of tools and content to support learning through video and other media elements. Importantly, however, large-scale empirical trials are still largely missing from this space. Ask almost any teacher or parent whether multimedia tools can support learning, particularly when accompanied by face-to-face and other instruction, and you are likely to get a resoundingly positive response. However, these claims should be validated, and perhaps more importantly, the specific mechanisms for the positive results seen in practice must be better understood.

In terms of technological advances in this space, improved video editing, crowdsourcing, and social sharing tools are all likely to enable the creation, collection, and distribution of media more easily. These advances should support teachers and parents in matching appropriate content to their students and children. Likewise, as software for playing multimedia elements improves, it should become possible to speed up or slow down content, pan and zoom to relevant elements for students with low vision, and generally customize the viewing experience depending on the needs of specific students. Likewise, these platforms should allow for greater collaboration and communication among professionals and parents, as individuals within a care network can collect and share relevant diagnostic and monitoring videos and other media.

CHAPTER 4

Mobile Applications

In this chapter, we overview the applications designed and developed for mobile devices to empower autistic people to learn new skills and conduct a variety of tasks (King et al., 2014). Although the lines between types and sizes of devices are continually being blurred, in this chapter we distinguish mobile devices from those primarily considered to be a desktop or laptop computer (as reviewed in Chapter 3), those intended for personal use such as wearables (as reviewed in Chapter 7), and those embedded in the environment (as reviewed in Chapter 8). The focus of this chapter is on devices that are primarily handheld, such as smartphones and tablets and can be used in multiple settings or anywhere the user goes. Because of this, the technologies have the potential to address many of the needs of the growing population of individuals with autism, their parents, and professional providers throughout their everyday lives.

4.1 OVERVIEW OF MOBILE APPLICATIONS

Most of the research and commercial applications in this space target preschool and school-age children, with limited interest in the adult population. This trend can be related, in part, to the overall emphasis on research for early intervention, and in part to the ability for older individuals to make use of non-autism–specific applications. In this chapter, we describe projects that make scholarly advances in these realms.

Autism Speaks (https://www.autismspeaks.org/), a leading advocacy and science organization dedicated to autism, and Autism Apps (https://www.autismapps.org.au/) provide a list of approximately 1,000 applications found in the Apple App Store or on Google Play Store (Figure 4.1). These marketplaces are rich with different applications, but in general, they are beyond the scope of this review unless they have been studied in the scientific literature.

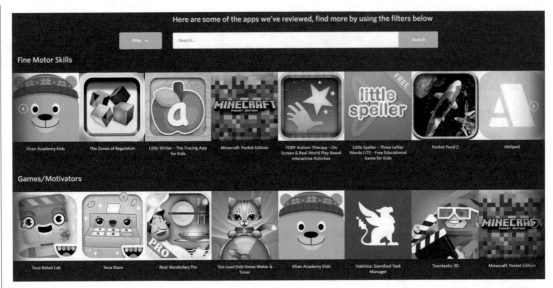

Figure 4.1: Autism Apps website showing a list of available apps reviewed for autism (August 2019).

4.2   MOBILE APPLICATIONS AND AUTISM

Since the launch of the Apple iPad, there has been a great deal of excitement in the autism community about collaborative multi-touch devices and their possible use in interventions for children with autism. The specific features of mobile devices, including low cost, portability, mobility, accessibility, robustness, durability, size, ease of recording, wireless internet access, and naturalness of touch interactions have facilitated the widespread implementation of this technology in schools, therapeutic centers, and houses.

Outside controlled settings, mobile technologies can be even more important for data capture. While many may be reluctant to instrument an entire house to record everything, caregivers still may have a need to capture and store information across a number of different locations and at different times. This kind of on-demand mobile data capture raises interesting questions about how to capture the right moments of interest. Kientz and Abowd (2009) examined this issue, relying on a host of persuasive techniques to encourage data capture as well as the equally important question of how to encourage review of these captured data. Mobile and ubiquitous computing tools have the potential to allow for review of data during casual moments of downtime and in a variety of environments. However, they still suffer from the challenges of getting people to review the data they have captured. Kientz (2012) highlights the way in which embedding access activities into capture activities can in some ways force reflection on the data, even if only momentary, a strategy repeated outside the autism world in personal informatics (Li et al., 2011). Smith et al. (2017) demonstrated how parents could be instructed to capture diagnostically meaningful videos of child behavior in the home using a smartphone. The video snippets were uploaded to a secure server where diagnosticians could then view and annotate them to inform a diagnostic evaluation.

Autistic children have benefited from mobile devices to compensate for limited verbal communication abilities, facilitate literacy development, increase overall academic performance, and decrease behaviors outside classroom norms (Giusti et al., 2011; Hourcade et al., 2013; Marco et al., 2013; Villafuerte et al., 2012; Zarin and Fallman, 2011). Mobile devices appear to be particularly appealing to parents ,too; studies have found that parents have high uptake of this technology and that 38% of the parents reported their children were using it for one to two hours per day (Clark et al., 2015).

Research has suggested that mobile devices are currently being used effectively with autistic children in a variety of distinct manners and for a variety of distinct functions (King et al., 2017), categorized below:

- the device functioning as a speech-generating and listening *Augmentative and Alternative Communication* (AAC) system (Lorah et al., 2015);

- the device serving as tool to deliver instructional video and video-based modeling (Burke et al., 2013) or exercise socio-emotional skills; and

- the device serving as a means to facilitate learning of academic and real-life contents (Chen, 2012).

We will use this categorization to review different mobile and tablet applications below.

4.2.1   MOBILE APPLICATIONS AND COMMUNICATION SKILLS

Several studies have explored the efficacy of software-based AAC for smartphones, tablets, and other multi-purpose AAC devices. The purpose of these studies was to investigate the utility of the mobile devices as a viable communication device by comparing the frequency of communication behaviors during conditions in which either a digital or paper-based system were used. The results were mixed and the findings were consistent in highlighting some increase in communication behaviors with the digital version, with no clear pattern across all students (de Leo and Leroy, 2008; Hayes et al., 2010; Canella-Malone, et al., 2009; Sigafoos et al., 2009; Son, 2006; Flores et al., 2012).

Proloquo2Go is the most studied app in the scientific panorama of digital applications aiming at improving communication skills. King et al. (2014) worked with 3 children with autism, aged 3–5, with limited to no vocal output to evaluate whether their requesting skills improved through use of the speech-generating capabilities. All participants achieved mastery, and the work supported the idea children could more quickly and efficiently to acquire communication skills via an AAC device, and even improve or acquire the ability of vocal requesting, regardless of technical difficulties and other barriers presented by these devices.

Waddington et al. (2014) studied Proloquo2Go as a means to make general and specific requests for access to toys with 3 autistic children aged 7–9. The study suggests participants improved and maintained their social communication skills using sequences, both with family and non-family members, and concludes that through systematized instructions, children with autism can take advantage of applications for mobile devices focused on communication support. Strasberger and Ferreri (2014) worked with 4 male children with autism aged 6–12 with limited or no vocal verbal behavior to investigate the acquisition of additional communicative behaviors. By using the device and the app, all children acquired the ability to request in a complete sentence; three of them could respond to "What do you want?" and two could respond to "What is your name?"

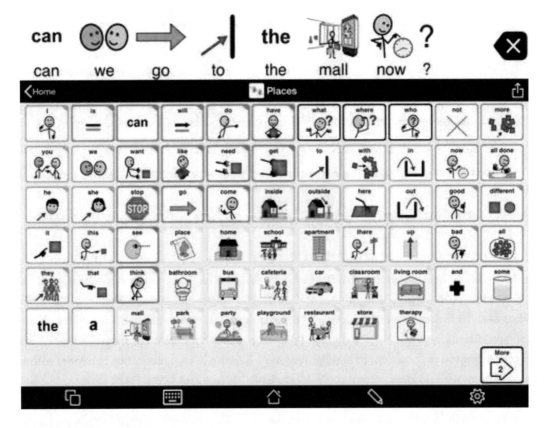

Figure 4.2: The Proloquo2Go app.

In general, most participants of these studies found mobile devices more appealing and, when questioned, caregivers reported a preference towards them. Only one child preferred the traditional picture exchange communication system (PECS) communication book (Ganz et al., 2013) and a teacher said he could not decide between the application or the traditional use of PECS (De Leo et al., 2011). Furthermore, advantages of the use of the mobile device included: dispensing with paper in traditional analog methods; the use of more eye-catching media and interactive stimuli; the accessibility and convenience of use; the ability to program, configure, and control; the possibility of creating safe learning environments; and the wide range of resources offered by both applications and mobile devices.

On the other hand, no studies demonstrated statistically significant advantages of the digital AAC system compared to traditional analog methods. Particularly, some studies pointed out differences between the two systems may offer benefits but cautioned that they may also detract from communication. Although the continual and rapid development of commercial products in this space has limited the need for development of research systems, it has not limited the need for empirical testing of these commercial systems, much of which is still lacking.

4.2.2   MOBILE APPLICATIONS AND SOCIO-EMOTIONAL SKILLS

The majority of mobile apps focus on the development of social and emotional skills. Regarding emotional skills, Sung et al. (2015) found most apps are intended to improve facial-processing abilities and facial expression recognition through naming expressions (e.g., Face Read 2), constructing expressions (e.g., Expressions for Autism, Let's Learn Emotions), or identifying them in social scenarios (Emotions 2 and Training Faces). Some apps targeted the recognition of facial identity by matching personalized faces to written names (Names to Faces) or text-to-voice (My Memory App) and one directly promotes eye contact by having the player identify a small face embedded in the iris region of the eye within a larger face (LOOK AT ME, Figure 4.3).

Gay et al. (2013) describe a mobile app called CaptureMyEmotion that enables children with autism to take photos, videos, or sounds, and at the same time senses their arousal level using a wireless sensor, then prompting them to comment on their emotion at the time of capture. Christinaki et. al (2014) found an increased acceptability using a natural user interface, such as the touch gesture of tablets, and that avatars helped in the learning process in the validation of an educational game for teaching emotion identification skills. Heni and Hamam (2016) designed the "World of Kids" app, which contains a set of games activated by the automated recognition of facial expressions, suggesting its potential for teaching emotions to children with autism who are inclined to want to play the games, as many children and adults are.

In general, most of the facial expression training apps ask the child to identify a target expression (e.g., smile as happy) across different people. People and their faces are depicted in different manners between the apps. Many apps use user-generated photos (Autimo, LOOK AT ME, Let's Learn Emotions, and Names to Faces) or stock photos of different emotions with faces from different ethnicities and ages to enable children to master their abilities in plausible scenarios and to train them on finding the same emotion across multiple individuals.

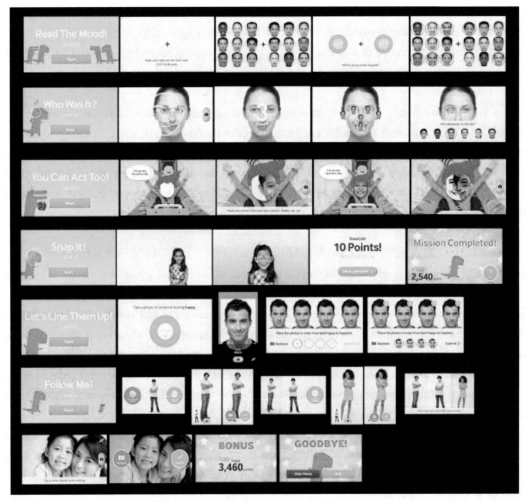

Figure 4.3: The LOOK AT ME App developed by Samsung©.

Other apps use cartoons or avatars to simulate expressions. This choice can only approximate real facial expressions but tends to be favored with so-called low- to medium-functioning autistic children. Cartoon faces are more focused on individual parts, such as the eyes, nose, and mouth, and are void of more detailed facial features, such as lines in the corners of a person's eyes when he/she smiles, to enable children to focus on the most important details.

In some apps, labeling of facial expressions is limited to the six basic emotions (happy, sad, angry, fearful, disgusted, and surprised), while other apps further built on more complex emotions such as tired or thoughtful (Sung et al., 2015). Finally, agents delivered through mobile internet-enabled devices have been shown in a study of 10 participants (7 with autism and 3 with general social phobia) to translate spoken phrases that were confusing or offensive into more easily understand-able language, thereby supporting learning of emotional communication (Bishop, 2003).

In other cases, custom technology developed specifically for social skills has been evaluated. Escobedo et al. (2012) present a seven-week deployment study of Mobile Social Compass (MO-SOCO), a mobile assistive app to help autistic children practice social skills in real-life situations, in which they might be discriminated against or ostracized for differing in their approaches to socialization. MOSOCO prompted students to practice social skills used in the classroom. In the limited deployment, researchers saw improvements in both quantity and quality of socialization for three autistic as well as nine neurotypical peers who used the prototype over a three-week period.

Figure 4.4: Students using MOSOCO (Escobedo et al., 2012) during recess (left) two students making eye contact (right).

Hourcade et al. (2013) also examined the use of tablet applications (retrieved from Open Autism Software[15]) to encourage social interaction for autistic students. The study compared after-school program activities conducted with apps to similar ones conducted without and found that, during the study, children spoke more sentences, had more verbal interactions, were more physically engaged with the activities, and made more supportive comments during activities conducted with two of the apps. The results suggest the approach to using the analyzed apps can increase positive social interactions in children with autism.

4.2.3   MOBILE APPLICATIONS FOR ACADEMIC, REAL-LIFE SKILLS, AND ASSESSMENT

The very nature of being available anytime and anywhere makes mobile devices particularly appealing for the support of learning activities and daily living tasks. Regarding learning, although mobile devices are useful in the classroom, both as AAC devices and beyond, learning does not just take place in classroom settings. In fact, generalization to settings outside those in which instruction is

---

[15]   http://homepage.divms.uiowa.edu/~hourcade/projects/asd/

initially provided is a huge challenge for education in general, and for autism in particular. A study from Karanfiller et al. (2017) aimed at teaching basic concepts to students who need special education and resulted in the design of a mobile application that will serve as a teaching aid. A usability evaluation found that a supervisor could enhance the teaching experience. Kraleva (2017) developed ChilDiBu, a picture-based app that combines graphic images, texts, and audio files to teach the Bulgarian alphabet, the numbers up to 20, and some basic colors. Eder et al. (2016) developed FillMeApp, an interactive mobile app to function as supplementary material to teach the human body schema and its parts. The researchers analyzed the results of the test survey and proved that the application is user friendly and interactive, while the logic of the game is understandable and motivational for learning.

Other researchers have explored how to support the different kinds of real-life practices required to generalize learning, particularly around non-academic topics, such as life skills (Tentori and Hayes, 2010). For example, students in workplace transition programs develop skills and confidence in the use of mobile technologies to reduce barriers to employment (Hayes et al., 2013). De Urturi et al. developed a set of serious games oriented towards first aid education, that is, how to handle specific situations and basic knowledge about healthcare, or medical specialties. Examples include HygieneHelper (Hayes and Hosaflook, 2013), VidCoach (Ulgado et al., 2013), and FirstAid (de Urturi et al., 2012). However, none of these tools has yet been evaluated. In a three-person study, Bereznak et al. (2012) demonstrated that mobile video modeling can also be used for self-prompting of vocational and daily living skills in persons with autism. However, this work indicated that mobile devices might require long-term rather than short-term use because they had to be returned in two of the three cases after a maintenance probe.

Prompting systems, when made mobile, can be moved into new environments. For example, Mechling et al. (2009) demonstrated that three student participants with autism could follow cooking recipes using a mobile device that adjusted prompting levels over time. Carlile et al. (2013) examined a variety of leisure skills in the home for four students, aged 8–12, and found that an iPod Touch–based activity schedule was effective in helping students stay on task and follow their schedules. However, these systems still require students to know when and how to use the system to remind themselves of their tasks.

Context-aware systems—systems that use information sensed from the physical surroundings to automatically adapt application behavior—can support these kinds of activities automatically (Rangel and Tentori, 2011). For instance, the COACH system uses environmental sensing coupled with a tablet-based interactive system to support hand washing (Bimbrahw et al., 2012). Although this prototype currently relies on heavy amounts of custom hardware mounted on site, one could imagine a future system in which the sensing architecture can dynamically connect to an individual's mobile device, be on the mobile device, or even worn on the body. The HANDS project, likewise, relied on artificial intelligence with sensed context to support teaching and development of social skills (Ranfelt et al., 2009).

One final interesting use of mobile devices has been in assisting with the assessment of aspects relating to autism. Baby Steps Text (Suh et al., 2016) is an automated text message-based screening tool that prompts parents to track and review developmental milestone data and connect them to resources using only text messages, with the intention of early identification of developmental delay including autism. This builds on previous work on Baby Steps as a software and Web interface to make it more accessible and easier to use, especially for caregivers who may not have access to data plans or internet at home. The tool was developed and evaluated with target families. During a one-month study, 13 out of 14 participants were able to learn and use the response structure (yielding 2.88% error rate) and complete a child development screener entirely via text messages. All post-study survey respondents agreed Baby Steps Text was understandable and easy to use, which was also confirmed through post-study interviews. Some survey respondents expressed liking Baby Steps Text because it was easy, quick, convenient to use, and delivered helpful, timely information.

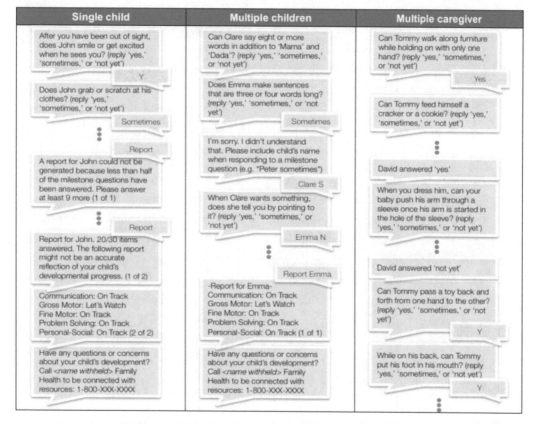

Figure 4.5: Example of Baby Steps Text syntax usage. Gray (left) indicates messages sent by Baby Steps and green (right) indicates messages sent by caregiver (Suh et al., 2016).

4.3   CLASSIFICATION SCHEME APPLIED TO MOBILE APPLICATIONS

In this section, we describe how two representative articles included in our classification system, Baby Steps Text (Suh et al., 2016) and MOSOCO (Escobedo et al., 2012), make use of mobile devices and tablets as a technology platform and fit into our classification scheme.

Baby Steps Text (Suh et al., 2016): The Baby Steps Text application uses text messaging to help parents complete early detection for the purposes of screening for developmental delay, including autism. It uses texting messaging and photos, and thus the input mechanism is *physical* and output mechanism is *visual*. Baby Steps screens using the Ages and Stages Questionnaire (Bricker et al., 1999), which covers the domains of *behavior, cognition, language/communication*, and *social/emotional skills* with the goals of *diagnosis/screening* and *parent/clinical training*. The target users are *family/caregivers* and clinicians and community organizations, who are *professionals*, and thus the settings for use include *community, home*, and the *clinic*. Because the system targets infants, the system was designed with involvement parents and professionals, but did not include autistic people themselves. Baby Steps Text was published at AMIA, which is a health informatics publishing venue in the medical field using a field trial *experiment*. It is being commercialized by the company Bright Steps (http://www.brightsteps.com), and thus it is now *publicly available*.

MOSOCO (Escobedo et al., 2012): The Mobile Social Compass (MOSOCO) system incorporates virtual and augmented reality into a mobile platform, which uses *spatial* and *physical* input and *audio* and *visual* output. The system was designed to support the generalization of skills learned through the Social Compass social skills curriculum (Boyd et al., 2013) and therefore targets the *social/emotional* skills domain. This paper describes a *quasi-experimental study* in which students with autism and their non-autistic peers used MOSOCO as part of an *educational intervention* in a *school* setting, primarily during recess and lunch. This paper was published in a *computing* venue, the ACM SIGCHI conference. The level of maturity was *functional prototype*, and the system required substantial researcher support to work in the uncontrolled environment of school free time.

More than any other platform in this book, mobile devices are showing an increasing tendency to be used in schools, homes, and other environments while simultaneously changing rapidly technologically and having limited empirical support for effectiveness of these interventions.

When we look across the research that has been conducted, however, there is immense promise for the potential of these technologies. As they become even less expensive and more ubiquitous, we expect the trend of developing autism-specific software as well as autism-relevant but non-specific software for these devices to expand. In light of this explosion, and the desire for insurance companies to reimburse families and schools to spend their money to adopt evidence-based practices, regulatory agencies, such as the Food and Drug Administration in the United States, are looking to require developers to support their claims—or change them—when they carry more than minimal risk to their users.

Although the development of the above regulatory strategies should spur additional empirical evidence and scientific research into the use of mobile tools for autism, the evidence base for the efficacy of these tools remains somewhat sparse, and this review cannot clearly substantiate whether smartphones and tablets have empirical evidence on autistic children's skills improvements. Most of the evaluations were exploratory, descriptive, or were conducted in one session; the findings on improvement of any skill are not conclusive. Accordingly, a recent study on mobile apps usability and accessibility shows most of the analyzed apps have limitations in the area of graphical layout and navigation (Dattolo and Luccio, 2016) and do not provide enough customization (Gelsomini, 2018).

Given the existing regulatory environment, many parents, clinicians, and educators may find themselves frustrated in using a low-cost device, such as an iPad or Android phone; having insurance regulations and acceptable-use policies may provide enough barriers that purchasing a custom, more expensive device, such as the DynaVox, becomes more feasible. However, we expect this attitude to change, much as initial regulations regarding laptops purchased for employees—only to be used for work activities—have gradually been reduced or eliminated as employers have begun to understand the unrealistic nature of purchasing and maintaining multiple multi-use devices for different functions.

Substantial work has already been published in computing venues, which are focused on more experimental technologies that are not yet ready for the mass market. Given the recent trend for researchers to develop prototypes that can be released to the mass market, through for example the Google Play store, this trend may change over time, with the lines blurring between research prototypes, commercially available product, and publicly available but free prototype tools.

Given the mass proliferation of mobile devices, there is no limit to the users and domains for which these tools might be applied. Thinking of mobile devices in terms of the personal computer of the 1980s or the internet-connected devices of the 1990s gives some indication of the size and scope of the mobile device space we expect to see going forward. Likewise, in our review, the technologies used already supported a wide range of labels in our categorization.

4.4   FUTURE DIRECTIONS

In autism, we have witnessed an evolution from low-tech strategies of the 1980s to custom high-tech strategies to multi-purpose hardware filled with "apps" that we see today (Shane et al., 2012). One look at the Google Play Marketplace or the iTunes App Store and a user will quickly become overwhelmed by the thousands of "apps for autism" and thousands more mobile apps not labeled explicitly for autism but potentially useful.

Detecting complex contextual information related to nuanced behavior is challenging, as is the secure archiving, transmission, and visualization of the data they collect. Few research projects have examined clinical or educational efficacy and the technological challenges of making robust, secure platforms and have shown a significant discrepancy between presumed benefit and actual clinical applicability or efficacy of apps (Gajecki et al., 2014; Heffner et al., 2015; Kertz et al., 2017). Despite numerous apps claiming to be evidence-based, this purported evidence often refers only to evidence-based principles such as cognitive behavioral therapy, which may or may not work equally well when delivered in a new digital format and app design (Fletcher-Watson et al., 2016). For example, in some individuals with ASD, skills learned through computer-based programs do not always translate into daily life (Fletcher-Watson et al., 2016; Whyte et al., 2015). Thus, understanding the evidence for available apps is an essential process for clinicians and families in making an informed decision regarding the use of a mobile device app in ASD. However, we expect this to change in the future, and the likely venues for publishing those results includes both autism-specific and other social science or health related venues.

The stage is set for the development of robust next-generation mobile apps as adoption rates have increased (Shanhong, 2019). Future developers should make use of the growing literature on gamification and serious games that promote the transfer of different skills in autism from the screen to the real world. In addition to advances on the theoretical front, there are several technological advancements that these training apps can exploit.

- Emotion-recognition software allows for real-time tracking of facial expressions and opens the doors for facial-expression-training apps that link facial emotions and everyday social situations. Whether these indeed detect "emotion" or simply a likely state that predicts some behavior is up for debate.

- Eye-tracking software is improving the ability of developers to track eye-gaze patterns—even away from computers—thereby enabling new game mechanics as well as additional surveillance and tracking mechanisms that could be used for health, consumer applications, or as part of policing. These approaches, of course, generate a variety of ethical considerations that must be addressed.

- User-generated content, in which the player takes photos and videos with the device's camera and incorporates the personally authored images into games, make a variety of systems both more customizable and more powerful in identifying, logging, and supporting everyday life.

- With adaptivity towards the user, systems will be able to adjust their behavior, considering the history of the users' interaction and their requests, needs, and preferences. Again, as in the other examples, this power will come with ethical questions about the tracking of people, particularly those who may be more vulnerable or marginalized, as a tradeoff for the potential benefit these approaches provide.

- Synthesizers will be similar (if not equal) to human voices, easing the recognition by users with limited comprehension, and supporting generalization to understanding real humans. So-called "deep fakes" have raised red flags about the proliferation of false information online, but these same technologies could be used to support autistic individuals with recognizing voices.

- Mobile devices can be connected to other different devices to work as a computational device or an interaction point. As an example, Google Glass or hearing aids, connected through a smartphone, allow for use of assistive and educational functionality alongside that of leisure and socialization to be delivered in more socially acceptable forms than previously obtainable.

While the resources and technologies are available, the challenge is for creative and innovative designers, developers, and other stakeholders to put their skills to the purpose of serving the autism community in cooperation with the autism community.

CHAPTER 5

Shared Interactive Surfaces

In this chapter, we discuss shared, interactive surfaces and their use in relation to autism. Shared interactive surfaces include applications that are intended for *multiple* users in a co-located, mostly synchronous, interaction, such as large displays, tabletop computers, electronic whiteboards, etc. However, with the popularity of larger touchscreen tablets that can be used collaboratively, we also review these as appropriate.

5.1 OVERVIEW

In the last decade, touchscreens have advanced far beyond their initial simple interaction capabilities. They can now accept multiple types of input, including from a finger or stylus and from multiple people at the same time, across a wide variety of sizes and form factors. Portable touchscreen devices, such as tablets and phablets (i.e., smartphones and tablets combined), allow for multi-touch interaction almost any time and anywhere. At the other end of the size spectrum, computationally enhanced tables (e.g., DiamondTouch (Dietz and Leigh, 2001), see Figures 6.3 and 6.4; Pixel-Sense[16]) allow face-to-face interaction and multiple simultaneous inputs from individuals acting independently or as part of a group (Morris et al., 2006). This may be especially appealing to people with motor or cognitive disabilities because the task of associating a physical device (e.g., a mouse) to a virtual representation can be difficult (Reed, P. (ed.), 1997; Whalen et al., 2006). In particular, the task of mapping the change in planes can be difficult cognitively, while the task of controlling the external device can be difficult for those with motor challenges.

These platforms can be useful in teaching social skills, such as turn taking, or in allowing for prompting and augmented learning from a peer or caregiver. These kinds of shared interactive surfaces are often inherently appealing to autistic people in much the same way as other shared interfaces, such as video games or even standard desktop computers. Their capabilities for enabling multiple people to interact simultaneously, however, takes them one step further in terms of encouraging interaction around and through a shared interface (Piper et al., 2006).

Because larger shared surfaces have had limited commercial availability and are fairly expensive to purchase, research into the use of shared interactive surfaces for autism is found more in technology-related publication venues than clinical or educational ones (Chen, 2012). However, as commercial products, these technologies have the potential to be used in a variety of clinical and educational settings, and the research surrounding their use is likely to expand as the cost of these devices decrease. In particular, this hardware enables researchers to ask important questions about the role these surfaces can play in supporting group-level interactions in therapeutic interventions, clinical encounters, decision making, and more. Not yet priced at a level that would allow for widespread home use, tabletops may lag behind tablets and phablets to some degree, but research conducted in this area is promising so far. Additionally, as larger-scale touchscreen televisions and similar screens continue to decrease in price and increase in distribution into the home and workplace, we expect to see more engagement in this space. The widespread use of interactive surfaces with larger screens, such as the iPad or iPad Pro, also make them more appealing for classroom use, including some examples of shared interaction between children, and between children and teachers. Although these technologies can be used for standard individual use or even non-co-located shared use, their real innovation lies in the ability to use them synchronously with multiple co-located people. Thus, it is this use case that is the focus of this chapter.

5.2   SHARED INTERACTIVE SURFACE TECHNOLOGIES AND AUTISM

Shared interactive surfaces have been used in a variety of ways, including the development of both academic and social skills. With most tablet and tabletop-based systems, users can engage in individual learning, thereby taking advantage of any of the kinds of programs described in the personal computer and mobile chapters (Chapters 3 and 4, respectively). However, given the unique capabilities for group interaction through these platforms, the majority of research in this area has focused on collaborative work and social skills, as we describe in this section, since, as in other areas, the predominant amount of work focused on children despite some promising results for adults.

Few studies to date have examined the differences in various hardware configurations, relying instead on feasibility and basic efficacy studies to determine the potential for these technologies. We expect this to be an area of much interest to autism researchers in the future, particularly those with an interest in social skills, visual supports, and technology. In one such study, Parenteau concluded that there is no conclusive recommendation as to whether to present stimuli vertically on a scan board or horizontally on a tabletop (Parenteau, 2011). Instead, based on a changing criterion study design with three students with autism, she recommends presenting stimuli in discrete trials to determine which presentation may improve acquisition rates. This study should be repeated with a larger sample to determine whether the lack of conclusive results stems from the limited sample size or whether other factors might be at play that could be used to determine appropriate approaches for these students.

5.2.1   LARGE, CO-LOCATED TOUCHSCREEN DISPLAYS

Although most shared interactive surface research projects focus on tablets and tabletop designs, some also make use of large touchscreens mounted vertically on desks or walls, such as the vSked project as well as a research effort focused on serious games for children with Pervasive Development Disorder-Not Otherwise Specified (PDD-NOS). In the multi-year vSked project (see Figure 6.1), researchers examined the replacement of paper-based systems for prompting, communication, teaching of academic skills, and token-based rewards in two classrooms (Cramer et al., 2011; Hayes et al., 2010; Hirano et al., 2010). This system included small tablet-sized interactive touchscreens (see discussion in Chapter 5) to be used by individual students at their desks and a shared active surface at the front of the classroom, primarily operated by the teacher but also at times by aides and students. Using an A-B-A study design, which began and ended with best practice paper-based tools, over several iterations of the prototype system, vSked was demonstrated to support students in learning both academic and social skills based on 202 hours of observation across 16 students, 2 teachers, and 8 aides in 2 classrooms (Hirano et al., 2010). Additionally, this project demonstrated that shared active surfaces could be used to improve communication among classroom staff as well as coordination and even friendly competition among students (Cramer et al., 2011).

Figure 5.1: vSked system uses a combination of wall-mounted displays and smaller tablets (Hirano et al., 2010).

In the Serious Games for PDD-NOS project, researchers at the University of Groningen and the Organisation for Applied Scientific Research in The Netherlands (TNO-NL) were interested in whether a vertical shared multi-touch surface could be used to teach both academic and social skills to students with PDD-NOS (van Veen et al., 2009). In this effort, six levels of mathematical problems were used to teach specific collaboration skills in a special-needs elementary school with 14 students (1 girl) aged 8–12. By playing the game for about 20 minutes per day, students saw improvements in collaboration within the game but were not able to show improvement in classroom skills.

The results of both the PDD-NOS games and vSked projects are promising. They indicate that shared active surfaces can be used in academic environments, such as classrooms. Indeed, other researchers have implored the community to think about games and other technologies within educational environments rather than specifically designed "educational technology" (Giusti et al., 2011). However, more work still needs to be done, including longer-term deployments to examine how well the technologies motivate students over time as well as how they might better support generalization outside of the program. Although the costs are quite high for these kinds of surfaces currently—particularly those with multi-touch capabilities and in the tabletop form factor—their prices are rapidly coming down. Additionally, with some work on the part of researchers or commercial product designers, existing SmartBoards, ever present and underused in so many classrooms, could be repurposed for these applications. Additionally, multi-touch capabilities are beginning to be incorporated into smaller form factors, as described in the following section.

5.2.2   COLLABORATIVE TOUCH-BASED INTERACTIONS

Increasingly, the Google Play Marketplace, Windows Marketplace, and iTunes App Store are filled with "apps for autism" for use on the increasing variety of tablets and phablets available in the commercial marketplace. As noted in the mobile-devices chapter, getting schools and insurance providers to pay for these multi-use devices—particularly when they have phone capabilities—can be challenging. As evidence of their efficacy increases, however, and as they become more commonplace, we expect this trend to change to some degree. Additionally, schools are currently struggling with acceptable-use policies in light of parental and student interest in using communication-enabled mobile devices in the classroom (Cramer and Hayes, 2010).

Tablets support multi-touch to a varying degree. Most can accept a variety of inputs. However, they do not typically support knowing who is producing which input, which can make developing for multiple users/players challenging. Despite this limitation, many commercial applications have been developed, and some research conducted, to support individuals with autism through these platforms.

For the most part, these efforts are described in the mobile chapter (Chapter 4). However, several research projects particularly stand out within this work focused on using collaborative tablets to teach social skills and is worth discussing as part of an examination of shared interactive surfaces. The first is the Open Autism Project (see Figure 5.2), where Hourcade and colleagues (2012) conducted participatory design with 26 students with autism, their teachers, and others concerned with the development of technologies for autism to create a suite of activities for multi-touch tablets (see Figure 5.2). The applications designed as part of this project were focused on encouraging social interactions "through creative, expressive, and collaborative activities" (Hourcade et al., 2013). In an A-B-A study of eight children (five who identified as boys and three who identified as girls at the time of the study) aged 10–14 with autism and low support needs, which began and ended with app use with custom-made matching paper-based activities in the middle condition, the researchers demonstrated that "children spoke more sentences, had more verbal interactions, and were more physically engaged with the activities when using the apps" (Hourcade et al., 2013). Although the majority of work in this chapter is dedicated to collaborative engagement with multi-touch active surfaces, the Open Autism Project shows the potential for individual engagement with these technologies alongside collaborative use. Thus, it is key to consider ways in which any given intervention may be developed for use under a variety of conditions, including but not limited to the number of students and facilitators required or allowed to engage with it at one time.

Application	Activities	Skills
Drawing	Collaborative storytelling and self-expression	Creativity, storytelling, fine motor skills, turn taking, sharing and collaborating, compromising one's interests with the interests of others
Music authoring	Collaborative and individual music composition	Creativity, fine motor skills, turn taking, sharing, and collaborating
Untangle	Visual puzzle solving	Talking aloud to cooperatively solve the puzzle, fine motor skills, turn taking
Photogoo	Emotion modeling	Understanding others' emotions, fine motor skills, detecting and predicting others' facial emotions

Figure 5.2: Hourcade et al.'s suite of applications for multi-touch tablets (Hourcade et al., 2012).

The second example of using interactive tablets as shared interactive surfaces in Sobel et al.'s *Incloodle* application (Sobel et al., 2016). This iPad app was designed to be used by two or more children of mixed cognitive abilities to encourage social-emotional development, empathy toward each other, and practice of inclusive educational practices. To promote inclusion, diverse characters describe their experiences, prompting children to ask each other questions, and children can take selfies together in specific prompted poses. The app uses the iPad's face recognition to ensure that both children are in the picture. In a lab study, Incloodle promoted inclusion between pairs of unacquainted neurodiverse children in which one child was typically developing and the other had a diagnostic label of autism or another closely related condition. Sobel also conducted a 10-week study in an inclusive kindergarten classroom and studied the ways that the system promoted inclusion among children over time (Sobel, 2018).

Figure 5.3: Sobel et al.'s *Incloodle* app where neurodiverse children collaboratively use an iPad to learn about each other, build empathy, and take pictures together (Sobel et al., 2016). The app uses the iPad's face recognition feature to ensure both children are in the picture (right).

Boyd et al. studied Zody, a collaborative iPad game to facilitate social skill development in children with autism aged 8 to 11 years. The Zody app had various cooperative features and helped children develop skills in turn taking, compromise, empathy, joint attention, communication, and shared enjoyment (Boyd et al., 2015). The Raketeer app (van Veen et al., 2009) used a single tablet in a two-player format to teach collaboration skills to children diagnosed with PDD-NOS, and the TabletG project (Winoto et al., 2016) guided two players at a Windows tablet to share an individual workspace for improving joint attention with autistic children in China. These examples show a strong desire for teachers and parents to use these apps for shared interactive experiences and the willingness to compromise screen space for portability and availability of platforms like collaborative tablets.

Figure 5.4: Two-player tablet interactions from Raketeer (van Veen et al., 2009) (left) and Tablet G (Winoto et al., 2016) projects (right).

5.2.3   USING TABLETOP INTERACTIONS TO DEVELOP AND PRACTICE SOCIAL SKILLS

By far the most published research in shared active surfaces for autism focuses on the use of tabletop interfaces in support of social skills for children, adolescents, and even adults with autism. These tabletop interfaces, often created explicitly to support Groupware (Grudin, 1994), naturally support a group of people working on them simultaneously or even together, enabling a wide variety of interesting interventions that would not be possible with other technological platforms.

One of the first projects in this area, SIDES (see Figure 5.4), involved a game collaboratively designed with a middle school social-skills class over several months (Piper et al., 2006). This game explicitly encouraged collaboration and decreased competition through a cooperative puzzle activity to be completed by four players on a DiamondTouch table that allowed the system to determine automatically whose turn it was and who was touching which locations on the interface. They tested their prototype game first with five students (all male, average age 12.6) from the same social-skills class with which they had been working, and then with two groups of four students of similar ages to the first evaluation (three students identified as boys and one as a girl), finding that games on these types of shared active surfaces can be motivating as well as effective in facilitating group work. Additionally, the authors note some key design considerations, many of which have been taken up by other projects described in the following paragraphs. In particular, they note that while identification of the users is helpful in enforcing game rules, the particular way in which the DiamondTouch identifies users—through tethered capacitive sensing—limits players in their physical interactions. Likewise, although they note that the students were largely able to use the system without any additional support, and in fact preferred system-enforced rules to those imposed by a human facilitator, the technology is still limited in its capabilities and requires an adult moderator to help the students process what they are experiencing through the game.

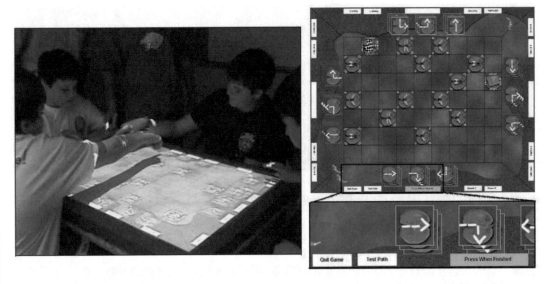

Figure 5.5: SIDES tabletop game uses the DiamondTouch platform to enforce turn-taking (Piper et al., 2006).

A relative explosion of tabletop games and interventions followed shortly after the SIDES project in the research literature. Most related to SIDES is a collaborative puzzle game (see Figure 5.5) built on the DiamondTouch platform (Battocchi et al., 2009) that was trialed with 70 typically developing boys and 16 boys with an autism diagnosis (Battocchi et al., 2010). In this work, the authors found that enforced collaboration had overall effects on collaboration and that the amount of coordination also increased for the children with autism through the use of an increased number of negotiation moves.

Figure 5.6: Collaborative puzzle game for DiamondTouch (Battocchi et al., 2009).

Originally developed for use by typically developing children during museum visits, the StoryTable system, also built on the DiamondTouch platform, was evaluated with 35 child dyads and found to encourage more complex language in stories and more evenly distribution participation in their creation (Zancanaro et al., 2006). The same research team then hypothesized that this interface might be useful to support children whose support needs are not severe. The team conducted a pilot intervention A-B-A study, in which six verbal autistic boys (in dyads) aged 9–11, used the StoryTable during eight sessions across three weeks to develop a collaboratively authored story (Bauminger et al., 2007; Gal et al., 2009). Based on a variety of outcome measures and analysis of the first 20 minutes of each session, the authors conclude that use of a co-located interface for these students can have positive effects on social interaction quality as well as quantity of repetitive and stereotypical behaviors. They further claim that the multi-user feature inherent to the DiamondTouch platform was particularly helpful in enforcing some tasks being done together, thereby improving the social skills measured experimentally. Others have built their own multi-touch hardware, such as the Par project (Silva et al., 2014), which included an application for understanding and promoting different collaboration patterns with autistic children with relatively low support needs.

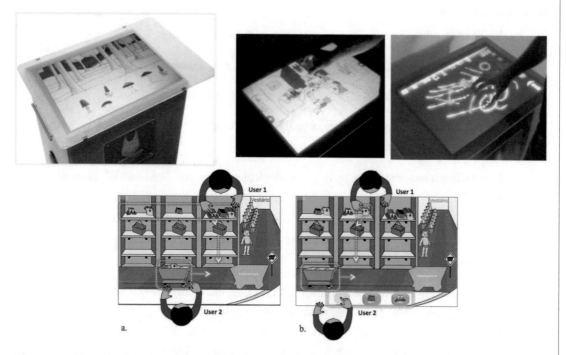

Figure 5.7: TrollSkogen project on an interactive tabletop surface (Zarin and Fallman, 2011) and the Par project (bottom) (Silva et al., 2014) on custom tabletop hardware.

Building on this concept of storytelling to support social-skills development, we now turn to the TrollSkogen project (Figure 5.6), in which the researchers use the concept of a Troll Forest, borrowed from Scandinavian fairy tales, as a platform for "micro-applications" designed to support learning of a variety of skills (Zarin and Fallman, 2011). The researchers evaluated this system with six children aged 5–8, all with diagnoses of either autism or Down Syndrome, finding that the micro-applications were helpful in allowing for some autonomy and choice by the children—namely which application to invoke at which time—while allowing for skill scaffolding and customization or expansion with ability over time.

In a similar effort, the Join-In suite of applications on a tabletop system supports teaching of social and other skills to autistic children (Weiss et al., 2011). In this work, the authors focused on three key dimensions to be supported by the design: joint performance of an action, sharing of personal resources to achieve a common goal, and planning together to coordinate actions and resources (Giusti et al., 2011). The authors deployed the applications with two therapists, who in turn used the system for social competence training with eight children with autism to determine whether the suite was usable and playable, involved the players in engaging and motivating experiences, and encouraged collaborative behavior. Based on this experience, the authors make several key recommendations.

- Move from educational games to games in educational settings.

- Address the game culture of today's children, which is always changing.

- Empower both the facilitator (Zancanaro et al., 2011) and the child.

- Switch easily between verbal, behavioral, and physical user interactions, a key finding in the SIDES paper as well (Piper et al., 2006).

- Ensure that elements for supporting ecological validity and "real-world" generalization exist.

5.3　CLASSIFICATION APPLIED TO SHARED INTERACTIVE SURFACES

In Figure 2.1, we reviewed four representative articles that use shared interactive surfaces as their primary technology platform. These included The Story Table (Gal et al., 2009), SIDES (Piper et al., 2006), vSked (Hirano et al., 2010), and Incloodle (Sobel et al., 2016). Here we describe how each of these fits into the classification scheme defined in Chapter 2, as well as discuss overall trends we observed for technologies making use of the shared interactive surface platform.

The Story Table (Bauminger et al., 2007): To support teaching collaboration and co-operation, prosocial behaviors, and language conversation and pragmatics, Bauminger et al. created The Story Table, which incorporates *physical* input and *audio* and *visual* output. This system supports the domain of *social and emotional skills* by creating an *educational intervention* through a tabletop display (in this case, specifically the Dia-mondTouch from Motorola). The authors previously conducted an *experiment* with 35 dyads using the system to facilitate complex and mature language (Zancanaro et al., 2006). The *functional prototype* uses a story metaphor and the overall concept of ladybugs traveling on the table in relation to the children's behaviors. In this paper, the authors used the system with three dyads of autistic children with low support needs to evaluate whether an intervention to support social skills could be developed around this technology. To our knowledge, autistic children were included as *users* in this project. This pilot intervention study used an A-B-A design, and the results indicated that the intervention can produce positive effects on some behavioral and communicative skills, particularly in the school setting in which the intervention was conducted. The work was published in a computing venue: the 6th Annual Workshop on Social Intelligence Design.

SIDES (Piper et al., 2006): SIDES is a four-player tabletop puzzle game *educational intervention* that supports children learning *social and emotional skills*, including in-creased collaboration and decreased competition. The primary input mechanism is *physical* and the output mechanisms are *audio* and *visual*. The authors evaluated the system in a *quasi-experimental* study with five students using a *functional prototype* from a social cognitive therapy class (all male, all with a diagnosis of some kind of neurodevelopmental disorder), and the system was built with autistic children as *informants*. Their results indicated that students were engaged with the system, but excitement around the technology itself sometimes created new behavioral challenges. This system, meant to be used by *children with autism*, can also be augmented with use by *clinicians/therapists* or *educators*, and, in fact, the authors found that some therapist coordination greatly improved the results of the system's use. This paper was published in a *computing* venue, CSCW, an Association for Computer Machinery (ACM) con-ference on Computer-Supported Cooperative Work.

vSked (Hirano et al., 2010): To support teaching collaboration and cooperation as well as language and academic skills, Hirano et al. developed vSked, an interactive visual scheduling and reinforcement system for use in classrooms. Its inputs are *physical*, via touchscreens, and its outputs are *audio* and *visual*. The vSked system was originally developed with the goal of *intervention/education*, specifically with a focus on visual scheduling, and includes both a *shared interactive surface* to show the overall schedule of the classroom and *mobile devices* for the students to use individually. The initial emphasis of the system on transitioning between activities, independently engaging in classroom tasks, and other *life/vocational skill*s, eventually gave way to the inclusion of *academic skills* as the teachers using the system adapted it to their needs. Additionally, the combination of a shared display with personal displays encouraged the development of *social/emotional skills*. The vSked system is used by *individuals with autism* and their *peers* in a *school* setting, facilitated by *educators*. This paper, describing a *quasi-experimental* deployment with one classroom, was published in a *computing* venue, the ACM SIGCHI conference. The level of maturity at the point of this publication was functional prototype, and autistic children were involved as *users* and *testers* in the design of vSked.

Incloodle (Sobel et al., 2016): Incloodle uses an iPad to encourage social skills and inclusion between neurodiverse and typically developing children with a selfie-taking app that encourages children to work together to understand differences and emotions. It uses the iPad's touch screen and the camera to ensure two faces are in the picture, and thus the input modalities are *physical* and *spatial*, and the outputs are *audio* and *visual*. The app is designed to encourage *communication*, *play*, and *social/emotional skills*, with the goal of an *educational intervention* between *individuals with autism* and their *peers* in a *school* setting. Incloodle was designed using participatory design with autistic children, and thus included them as *informants* and *design partners*. Incloodle was published at a computing venue (CHI 2016) and evaluated in an *experimental* setup as a *functional prototype*.

The advent of easily programmable tabletop platforms, like DiamondTouch and eventually Microsoft Surface—now known as PixelSense—greatly expanded interest in research around shared active surfaces. Likewise, other interaction paradigms, such as those afforded on walls and other surfaces by large displays—Kinect, the Wii, and other systems—has expanded the definition of these surfaces. However, they are still very expensive to buy commercially and challenging technologically to create. Thus, the work is still fairly preliminary. As these products come down in price, and people use more small-scale interactive surfaces, such as larger-sized tablets, we expect to see more work in this area. The majority of shared interactive surface work in our review was focused, perhaps unsurprisingly, on social skills, and also due to the more novel computing platforms and techniques required, were primarily focused on computing venues. These platforms provide a compelling set of functionalities for exploring social skills, particularly when they allow multiple users to interact with them at once.

5.4   FUTURE DIRECTIONS

Similarities in interaction between shared interactive surfaces and the currently used best practice of paper-based systems enable transfer of a wide variety of interventions and approaches to this platform with relative ease. Of course, the interactive and computational features of these platforms also allow for the development of new interventions. Most of this research has so far focused on establishing the feasibility of such approaches. However, initial efforts have investigated the efficacy of these approaches in accomplishing the goals set forth, particularly for large co-located touch-screens, multi-touch tablets, and tabletop interfaces. However, there is room to go beyond this work through larger-scale clinical trials or evaluation of the specific features of the platforms themselves as described above, or through expansion into other application areas and increasing technological innovation as described in the next section. Given the expense of the systems, especially tabletop platforms like DiamondTouch and Microsoft PixelSense, these may still be a way off. As tabletop, multi-touch, and other shared active surface technologies become more commonplace, it is likely that we will see a surge in both research and commercial applications in this space. In particular, the placement of these systems in homes, schools, and clinics for the therapeutic interventions described above may open the door for use of the hardware for other purposes. For example, one could easily imagine the Abaris or Walden systems described in Chapter 3 making use of shared interactive surfaces to support therapy as well as recordkeeping rather than piecemeal solutions of computationally enhanced pens and paper, tablets, video capture, and physical artifacts (Hayes et al., 2004). Less expensive and more portable and versatile shared surfaces, such as larger-sized iPads that can be shared between children, may see more use in practice sooner.

Another area of expansion is in the development of multi-sensory environments and virtual reality environments that include shared interactive surfaces, covered in (Chapter 6), as well as sensors (Chapter 7), mobile technologies (Chapter 4), and other elements covered in detail elsewhere in this book. One research project has begun in this direction, MEDIATE. This system is an immersive physical-digital environment that is highly dependent on a variety of shared active surfaces for the floor and walls. Intentionally neither "therapeutic nor educational," the goal of MEDIATE is to let people with no verbal skills "express themselves" and have fun. The floor surface reacts to footsteps by generating sound, the screen walls react to movement and touch, and so on (Parés et al. 2004, 2005a). We discuss MEDIATE in greater detail in Chapter 6.

Shared interactive surfaces have been demonstrated repeatedly to be easy to use for people with a wide variety of abilities, including low vision, low motor skills, and other physical disabilities.

The ability to use them collaboratively is encouraging and exciting for the development of a variety of peer- and facilitator-based interventions. Additionally, beyond the need for, and likelihood of, larger and more complex trials of these technologies, we predict that a variety of innovative uses both independently and in concert with other technologies will be seen in the near future.

CHAPTER 6

Virtual, Augmented, and Mixed Reality

In this chapter, we describe a brief overview of work involving virtual (VR), augmented (AR), and mixed (MR) reality with autistic people. For the purposes of this work and using our technology platform classification scheme, this chapter includes only VR, AR, and MR applications, eventually focusing on those that use head-mounted displays, often called visors, viewers, or headsets. Examples of viewers available at the time of writing are Google Cardboard and Daydream, Samsung Gear, Facebook Oculus Rift, HTC Vive, Sony PlayStation VR, Microsoft Hololens, andMagic Leap.[17]

6.1 OVERVIEW

There are different scientific and marketing definitions about virtual, augmented, and mixed reality. For the purposes of this book, however, we define them below to reduce confusion and aid in the readability of this review.

The suite of technologies and techniques generally referenced to as virtual reality (VR) uses computational technology—including sensors, headsets, and input devices—to create a simulated environment. Frequently, in these environments, users are represented by their digital alter-egos, or avatars. In multi-user VR environments, these avatars can interact with one another and/or with simulated virtual users, often referenced as bots or sims. VR employs a variety of digital displays to share the experience with the user, including head-mounted as well as more traditional screen-based designs.

Augmented reality (AR) shares the use of computational technology to create a simulated environment. However, it is distinct from what is traditionally called VR in that it "augments" or overlays "reality," that is, the physical world, with digital output. One famous and early example of augmented reality includes the ever-present stripes on the televised field during American football games showing viewers the line of scrimmage or first-down marker. In the mobile arena, the most well-known example of augmented reality is likely the Pokemon Go game, which took the world by storm in the second decade of the 21st century. In this game, people use their mobile phones to view and "catch" geo-located Pokemon characters in the physical world. A wide variety of heads-up

---

[17] Google Cardboard and Daydream (https://vr.google.com/), **Samsung Gear** (https://www.samsung.com/it/wearables/gear-vr-r323/), **Facebook Oculus Rift** (https://www.oculus.com/), **HTC Vive** (https://www.vive.com/us/), **Sony PlayStation VR** (https://www.playstation.com/en-us/explore/playstation-vr/), **Microsoft Hololens** (https://www.microsoft.com/en-us/hololens), **Magic Leap** (https://www.magicleap.com/)

displays, eyeglass-mounted displays, and mobile tablets and phones serve as the platform for this increasingly prevalent form of technological engagement.

Over the last 30 years, researchers have attempted to describe and categorize virtual reality according to a wide variety of metrics. Indeed, as the hype of each new iteration of VR hit its crescendo, a series of descriptions of these technologies would emerge alongside, each promising to fulfill the promise of VR once and for all. Despite these wide-ranging categorizations, however, it is still often useful to consider Milgram's simple spectrum from the "real" to the "virtual" (see Figure 6.1) established in 1995 (Milgram et al.,1995).

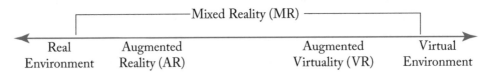

Figure 6.1: Virtuality continuum (Milgram et al., 1995)

However, it is also notable that this perspective often creates a false dichotomy and neglects the very real physicality of wearing a head-mounted display, experiencing the motion sickness of engaging in both physical and virtual realms simultaneously, or even just the pure fact that all code is written out to, and runs on, physical infrastructure of memory, wires, power sources, and processors. Thus, we caution readers to recognize both the utility and the oversimplification of Milgram's model (see Figure 6.1).

Defining experiential features in these realities include interactivity, immersion—the degree to which a virtual reality user feels engrossed or enveloped—and presence—the degree of feeling situated within the virtual environment (Burdea and Coiffet, 2003). Similarly, Azuma (1997) states that augmented reality "enhances a user's perception of and interaction with the real world: the virtual objects display information that the user cannot directly detect with his own senses. The information conveyed by the virtual objects helps a user perform real-world tasks." These platforms have the potential to improve and support of a variety of behaviors and experiences, such as concentration, socialization and communication, self-awareness, and memory and cognitive functioning (Guazzaroni and Pilliai, 2019).

6.2   VIRTUAL, AUGMENTED, AND MIXED REALITY AND AUTISM

In our review for this chapter, we organize the technologies reviewed into the following categories: (1) virtual reality with headsets and (2) augmented and mixed reality.

6.2.1   VIRTUAL REALITY WITH HEADSETS

The application of VR technologies for use by individuals on the autism spectrum began in the 1990s through the pioneering work of Dorothy Strickland (Strickland, 1996, 1998; Strickland, et al., 1996) and Cheryl Trepagnier (Trepagnier, 1999). Since their incipient efforts, and the subsequent work of others (e.g. Parsons, Garzotto, Gelsomini, and Bozgeyikli), it has been suggested that VR is an especially useful technological medium for individuals on the autism spectrum in preparation for, as an adjunct to, and/or in place of learning concepts and skills in real-world environments. The myriad putative affordances associated with virtual reality technologies for autism that have been suggested (as reviewed in Parsons and Mitchell, 2002; Parsons, et al., 2004, 2013; Rizzo, et al., 2006; Vera, et al., 2007; Bellani et al., 2011; Gelsomini, et al., 2016; Garzotto, et al., 2017b) include the following:

- They are programmable spaces that can be carefully selected and controlled, thereby enabling tailoring of content to meet individual needs.

- Their virtual elements and properties can be altered in real time.

- They permit access to inaccessible environments that are unsafe until basic skills are learned.

- They are repeatable and facilitate practicing skills across a range of contexts and periods of time.

- The combination of their adaptability and their exact replication make them ideal clinical test grounds for assessments and behavioral tests (Duffield et al., 2018).

- They circumvent face-to-face interactions and require no or minimized human interaction, which might be uncomfortable and/or overwhelming while learning new skills.

- They remove competing and confusing stimuli from the social and environmental context.

- The interactivity they facilitate is entertaining and can increase motivation to learn and practice skills.

- They are mainly visual, and people with diagnoses along the autism spectrum are thought to benefit from experiencing improved learning outcomes when presented with visual information.

- They can shape and visualize abstract concepts, necessary for children with autism.

- They reduce the cost of traveling to some environments repeatedly.

- They are realistic and support role-playing and active participation within representative settings from different perspectives.

- They create alternative realistic scenarios that facilitate easier generalization of concepts by variation.

- They reduce environmental complexity and simplify stimuli to enhance salience.

- They can be paused, permitting an educator or researcher to narrate what is being seen, what to attend to, and how to respond.

- Their verisimilitude supports learning that increases the probability of skills transferring to real-world environments.

- They represent a safe environment to make mistakes, incrementally building confidence in users before they apply skills in the natural environment.

- Autistic people frequently learn how to use and interact with VEs and avatars quickly and show significant improvements in a few trials.

- While not systematically tested, it has been suggested that VR is well suited for determining whether a variety of psychological theories (theory of mind, executive function, weak central coherence, embodied cognition, etc.) are explanatory for behavior in individuals with autism (Rajendran, 2013). As these theories are tested over time and new ones emerge, VR may be a helpful testbed for checking the efficacy and appropriateness of such theories for autism.

- Despite substantial potential benefits, the risks to current VR technologies can be of great concern to healthcare professionals (Markopoulus, 2018), teachers, parents, and autistic people.

We next describe the study of virtual environments in terms of access and presence; sensory perceptual functioning; and assess and/or instruction related to safety skills, selective and sustained attention, and social skills.

Figure 6.2: Experiences in the Magic Room: A-B. Relaxation with sea-world projections and space, C. Magic Ball, D. Pass the light, E. Shapes, F. Storytelling (Garzotto and Gelsomini, 2018)

Presence, Immersion, and Acceptability to Virtual Environments

A number of studies have found that autistic people experience high levels of immersion and presence, successfully navigate virtual environments (VEs), can track stimuli presented within them (Brown, et al., 1997; Charitos, et al., 2000; Eynon, 1997; Kijima, et al., 1994; Mineo, et al., 2009; Parsons et al., 2004; Strickland, 1998; Strickland et al., 1996; Wallace, et al., 2010). Some studies also explore the acceptability of VR head-mounted displays and their side effects such as cyber-sickness (Parsons, et al., 2005; Wang and Reid, 2011; Holden, 2005), finding that performing familiarization exercises at the beginning of study can lower barriers and enhance acceptability (Gelsomini, et al., 2016). Other reviews have demonstrated the challenges and opportunities to use of virtual reality technologies by people with autism (Bozgeyikli et al. 2018a; Bradley and Newbutt 2018; Mesa-Gresa et al. 2018; Turnacioglu et al., 2019).

Far from the "one size fits all" solution that some industry leaders would have us believe them to be, virtual environments are complex technologies with immense promise but that require specific attention in their design to the cognitive differences that often coincide with autism (Boyd, 2018; Boyd et al., 2018a; Bozgeyikli, et al., 2018b). In response to these design challenges, Newbutt and Cobb (2018) proposed a framework for implementation of virtual reality technologies in schools for autistic students, which is an extension of a prior framework developed for virtual reality in schools more broadly (Crosier et al., 2002). The three key elements of design considered in that work include "facilities and equipment available, intended use in school and individual learner characteristics" (Crosier et al., 2002). These primary design considerations should be carried forward both in schools and out and are considered in our assessment of research applications in the following sections.

Figure 6.3: A participant experiencing the street environment inside the Blue Room (Wallace et al., 2010).

VR for Social and Emotional Development

Given the heavy emphasis on social skills development for autistic children, it is perhaps not surprising that most VR studies with autistic people found in our review focus on this area. Some applications focus broadly on teaching social skills generally (Neale, et al., 2002; Parsons, et al., 2006, 2004, Bekele, et al., 2016), while others are more focused on specific considerations, such as symbolic play (Herrera, et al., 2008), social cognition (Michelle, et al., 2013; Didehbani, et al., 2016; Yang et al., 2018), social conventions (Mitchell, et al., 2007; Parsons, et al., 2005), conversation skills (Politis, 2019), social understanding (Cheng, et al., 2015), social adaptation (Ip et. al, 2018), virtual peer and agent interaction (Burke, et al. 2018; Cobb et al., 2002; Tartaro and Cassell, 2008), and peer-to-peer VR-mediated interaction (Bauminger, et al., 2007; Loiacono, et. al,, 2018), specifically.

Figure 6.4: An example of co-presence inside Social MatchUP (Loiacono, et al., 2018)

Collaborative virtual environments can be particularly useful for socialization, social support, and social skills development. For example, Hand-in-Hand focuses on supporting children interacting socially in a collaborative virtual space (Zhao et al., 2018a). As another example, Ghanouni et al., (2019b) developed a platform to use social stories within a virtual reality system. The general approach of social stories for supporting people with autism to think through, practice, and reflect on social situations gains additional richness and engagement when used in this type of platform. Similarly, though not a fully immersive virtual reality platform, Ringland et al. (2016) found a variety of naturally emergent social engagements in Autcraft, a Minecraft server dedicated to autistic children and their allies.

One critique of virtual reality—and part of why its adoption is limited in clinical and educational settings—relies on the "realness" of the interaction (Markopoulus, 2018). We can debate heavily what constitutes real, and indeed Kate Ringland's dissertation covers much of this debate (Ringland, 2018). However, one approach to dealing with these challenges is to make the virtual environment more responsive to the physical. For example, Babu et al. (2018) developed a social virtual reality world which is sensitive to gaze. Similarly, to support the Hand-in-Hand system mentioned above, Zhao et al. (2018b) developed a haptic-gripper system that could also be used for assessment alongside therapeutic engagement. Some recent studies have combined eye-tracking technology with VEs to assess gaze patterns to peer avatars as an index of social attention abilities (Alcorn et al., 2011; Jarrold et al., 2013; Lahiri, et al., 2011). Findings indicate that this paradigm can both distinguish individuals with autism from those without and enhance social attention through virtual manipulation.

Similar to other technology platforms, the recognition of emotions in others and the acquisition of social skills has also been an important application area. A number of studies have demonstrated that collaborative virtual environments (CVEs) (Cheng and Ye, 2010; Fabri and Moore, 2005; Moore, et al., 2005; Loiacono, et al., 2018) and avatars (Fabri, et al., 2007; Golan et al., 2010; Hopkins et al., 2011; Mower, et al., 2011) are useful to assess and teach individuals with autism to recognize, understand, judge, and appropriately respond to facial expressions, which some also equate to emotions, while others disagree that this kind of technology can actually engage emotion as we understand it.

Support for Work and Autonomy

Support for people who identify as having autism to live and work in a way that preserves agency and autonomy is a substantial goal for many people. One major thread of research has been teaching individuals with autism different skills in a way that is safer and less threatening than a "real world" experience might provide. For example, studies have demonstrated that a variety of safety skills can be successfully taught using VEs, including crossing the street (Josman, et al., 2008), responding to fires (Strickland, et al., 2007) and tornado warnings (Self, et al., 2007), driving (Wade, et al., 2016), shopping in a supermarket (Adjorlu, et al., 2017; Lamash, et al., 2017; Fridhi, et al., 2018; Garzotto, et al., 2017b), and going to a museum (Gelsomini, et al., 2017; Garzotto, et al., 2017b). In addition to safety skills and other experiences that can be taught to anyone, Maskey et al. (2019a, 2019b) found that virtual reality could be used to support autistic youth in overcoming fears and phobias.

Figure 6.5: Example of VR-based social stories in a museum and in a supermarket (Garzotto, et al., 2017b; Gelsomini, et al., 2017)

Beyond simple everyday activities, however, many people with autism seek to work, with the goal of many families, rehabilitation centers, and other government agencies focused on full-time, paid employment. One of the major barriers to work is transportation. Thus, a virtual reality platform for teaching driving skills is particularly promising (Bian et al., 2019). Similarly, improving performance in workplace interviews (Arter et al., 2018) can make a substantial difference in the lives of people seeking self-sufficiency.

Once employed, barriers to continued employment can often include cognitive and self-regulation differences. Thus, VR has also been explored to support cognitive training. For example, Gelsomini, Garzotto, and Occhiuto (2016, 2017) developed Wildcard, which supported the development of selective attention—the capability to focus on an important stimulus, ignoring competing distractions—and sustained attention—the capability to hold attention for the time needed to conclude a task. Using this platform, the researchers found an improvement in attention both in the VR application and on paper-based tests.

Virtual Physical Activity

Although virtual reality by definition includes the engagement within a virtual space, applications for physical activity utilized either real walking or standard controllers such as gamepads, joysticks or keyboards to support feelings of physical movement within these virtual spaces. Real walking was used in different forms; some studies used electromagnetic tracking in a Cave Automatic Virtual Environment (Finkelstein, et al., 2010; Cai, et al., 2013), while some studies used Microsoft Kinect for tracking (Bartoli, et al., 2013; Garzotto, et al., 2014), but without the use of VR headsets. One study has evaluated the use of fully immersive VR to assess visual and vestibular functioning as it relates to postural reactivity in autistic people (Greffou, et al., 2012). To move inside the VEs, Bozgeyikli et al. (2016a) implemented three commonly used locomotion techniques (redirected walking, walk-in-place, and joystick controller), two unexplored locomotion techniques (stepper machine and "point and teleport"), and designed three locomotion techniques (flying, flapping and trackball controller).

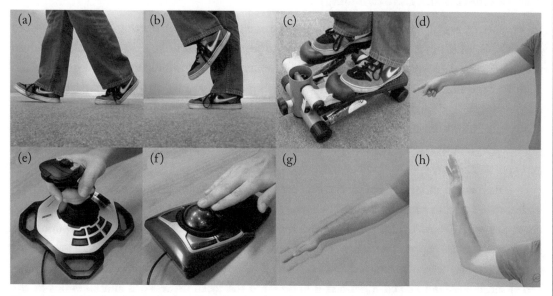

Figure 6.6: Representative photos of the locomotion techniques used in Bozgeyikli et al. (2016): (a) redirected walking, (b) walk-in-place, (c) stepper machine, (d) point and teleport, (e) joystick, (f) trackball, (g) hand flapping, and (h) flying.

Beyond the simple movements of walking and controlling virtual physical objects, virtual environments can also be used for more elaborate physical movements. For example, both Dance-

Craft (Ringland et al., 2019) and Choregrafish (Altizer et al., 2018) use a mix of physical and digital "realities" to support autistic people participating in dance therapies.

6.2.2   AUGMENTED AND MIXED REALITY

Augmented and mixed reality sit further on the spectrum towards physical engagement than virtual reality. The integration of a variety of input and output devices enables innovative ways to engage with the technology, such as gesture-based control and spatial mapping. To facilitate these kinds of interactions, augmented reality systems often require substantial sensing systems and inference engines, which can also be repurposed in clinical applications to collect quantitative data for clinical research (Liu et al., 2017). In terms of therapeutic benefit, these systems can be useful for scaffolding generalization of skills learned in a virtual world to the physical world and can allow individuals with autism to receive additional help while engaging in everyday activities.

Just as in virtual reality, augmented reality applications include social skills and communication support, cognitive skills development, and physical engagement. However, overall, this space is less developed than virtual environments, due in part to the newness of AR platforms.

Social Support, Skills, and Scaffolding

In support of social engagement, Escobedo and colleagues (2012) developed a system called MOSOCO, which uses AR to prompt students to practice social skills used in the classroom. In the three-week deployment of the prototype, researchers saw improvements in both quantity and quality of socialization for three students with autism as well as nine neurotypical peers. Bai and colleagues (2012) developed an AR tool with the goal of helping comprehension and flexibility of object substitution during pretend play. Their system is still a proof of concept that has not yet been evaluated rigorously but has received positive feedback from autism domain experts on its feasibility and appropriateness.

Augmented and mixed reality can support children in becoming more aware of their physiological state and thus adapt their social behavior (Mcduff, et al., 2017). Chen and colleagues (2016) developed an AR-based video modeling storybook to improve autistic children's perceptions and judgments of facial expressions and emotions. Results showed the system, providing an augmented visual indicator, effectively attracted and maintained the attention of children with autism to nonverbal social cues and helped them better understand the facial expressions and emotions of the storybook characters.

Other researchers based their AR applications and interaction on external devices such as Microsoft Kinect, Leap Motion, and Smart Glasses. For example, the DanceCraft system noted above relied on Kinect once developed in its AR form (Ringland et al., 2019). Syahputra and colleagues (2018) developed a 3D animation of social story by detecting AR markers located in special book and some simple games based on leap motion controller to read hand movement in real-time. The system was tested with three autistic people and showed it can support visualizing social story therapy in interpreting and understanding the surrounding social situations.

Figure 6.7: 3D animation of social story (Syahputra et al., 2018).

Casas and colleagues (2012) have used Microsoft's Kinect to combine the virtual and real worlds. Casas and colleagues' system created what they call a Pictogram Room on a screen, which superimposes images over a live video to teach individuals different skills, such as body awareness and interacting with others. Their evaluation with five students with autism mostly focused on testing the feasibility of the method, with three of the students successful in carrying out tasks with the system. Savvides et al. (2010) similarly developed a mixed-reality game to support communication for students with autism, which could be used by autistic children in working together to communicate.

As wearable interfaces have become more common commercially as well as simpler to prototype in labs, additional AR approaches have begun to be explored. For example, Boyd et al. (2016) used the commercially available Google Glass platform to provide cues about volume and tone of voice in face-to-face conversations, finding autistic people were able to modulate volume with feedback but faced greater challenges regarding tone. Similarly, the wearable system ProCom was useful for supporting people with autism in self-modulating distances from others to match norms (Boyd et al., 2017; Jiang et al., 2016).

Educational and Cognitive Interventions

Augmented reality, far more so than truly immersive virtual environments, can be used less obtrusively in educational settings. Thus, some of the most promising interventions and prosthetic uses of AR are in the space of educational and cognitive interventions.

Keshav et al. (2018) developed a smart-glasses intervention to aid children with attention and social educational learning. This intervention was found to be logistically practical to implement, easily usable by both the educator and student, and not time-consuming to learn. Educators identified the experience as being fun for the student and felt that the student demonstrated improvement in his or her verbal and non-verbal skills. There were no adverse effects on the other students or the classroom, and the technology did not result in a distraction. These findings suggest that interventions delivered by smart glasses may be practical, useful, and lead to improvements in skill development.

Similarly, Lumbreras, et al. (2018) developed Aura to support learning in classrooms. This system was notably tested with very young students (aged 4–8), much younger than most work in augmented reality. Even at this young age, the autistic students were able to use the mobile phone-based AR system to complete 42 activities across five modules, including learning shapes, drawing, and more emotional-based skills, such as empathy and emotion recognition.

Figure 6.8: (a) The participant and facilitator converse prior to the intervention while sitting opposite each other in a classroom setting. The participant displays minimal visual attention to facilitator and displays fidgeting behavior with his hands. (b) The facilitator prepares the student for the intervention, communicating via verbal and non-verbal cues. The participant continues to display minimal visual attention towards the facilitator, instead focusing on his hands and wrists. (c) The participant now wears the smart glasses and receives intervention while interacting with the facilitator. The participant looks directly at the facilitator during the intervention, and no longer displays hand-based fidgeting (Keshav, et al., 2018)

Physical Engagement and Support for Autonomy

Just as AR can be used in educational settings, it is also promising in its ability to work as a support in the workplace and other everyday contexts. For everyday use, however, the platforms are still somewhat challenging. As an example, Microsoft Hololens's manual identifies the potential side effects as nausea, motion sickness, dizziness, disorientation, headache, fatigue, eye strain, dry eyes, and seizures (Sahin, et al., 2018), although their occurrence among users with autism has not been studied.

Despite these risks, everyday uses, such as navigation and transportation support, may be compelling enough to warrant overcoming these challenges. For example, McMahon and colleagues (2015) examined the effects of location-based augmented reality navigation compared to Google Maps and paper maps as navigation aids for students on the spectrum. The study measured their ability to independently make navigation decisions in order to travel to unknown business locations in a city and found that students traveled more successfully using augmented reality.

Aruanno and colleagues (2018) developed HoloLearn, an MR application that provides enriched learning and training experiences on simple everyday tasks in domestic environments and aims at improving autonomy. HoloLearn was well accepted by the participants and the activities in the MR space were perceived as enjoyable, despite some usability problems associated to HoloLens interaction mechanism.

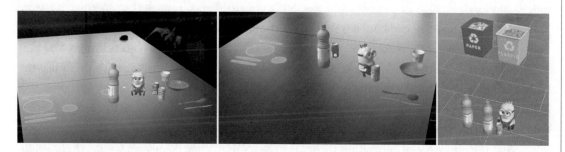

Figure 6.9: Aruanno et al. (2018). a virtual minion-like assistant who helps set the table and correctly sort recyclables.

Recently, Sahin et al. (2018) developed an AR system based on Google Glass called Empowered Brain to evaluate negative effects and design concerns regarding this technology. Results demonstrated most users were able to wear and use Empowered Brain, with most reporting no negative effects and no design concerns.

6.2.3   MULTISENSORY VIRTUAL ENVIRONMENTS

Multisensory virtual environments have been used for many years in support of educational and therapeutic interventions for people with a wide variety of disabilities, such as multiple learning difficulties (Mount and Cavet), dementia (Baker et al., 2001), intellectual disabilities and autism (Kaplan et al., 2006), blindness (Lahav and Mioduser, 2008), and traumatic brain injury (Hotz et al., 2006). The exact approaches of each multisensory environment can vary and, in some cases, have been tested against one another (e.g., Fava and Strauss 2010). However, some commonalities can be found. Typically, a multisensory environment includes a physical space equipped with items that provide stimulation of different senses within a nonthreatening environment (Shams and Sietz, 2008). Their designs are grounded in the theories of sensory integration (Bundy et al., 2002; Ayres, 1972) and can also include virtual elements. Learning within these environments focuses on processing the experienced sensory information and integrating that input to develop cognitive maps, and to plan and organize behavior.

Differences in neurological processing and integration of sensory inputs can result in challenges with completing certain tasks or performance of socially adaptive behaviors. Most often clinically referenced as either sensory processing disorder or sensory dysfunction, these differences in sensory systems can make learning in traditional environments difficult and create challenges with performing everyday tasks in environments with sensory stimuli that are non-optimal for each individual's sensory systems. Sensory-based treatments are based on sensory integration theory to provide individualized, controlled sensory experiences to help modulate responses to environmental input (Boyd, et al., 2010). These treatments use a variety of sensory modalities (e.g., vestibular, touch, auditory), a range of passive (e.g., wearing a weighted vest, massage) to more active (e.g., jumping on a trampoline, climbing a wall) activities, and target hyper or hyposensitivities.

Sensory integration theory posits that sensory integration is a microbiological process that organizes sensation from one's own body and the environment and makes it possible to use the body effectively within the environment (Thompson, 2011). The theory of sensory integration is based on cognitive plasticity, sensory integration, and the notion of a hierarchical brain organization that ties these all together (Ayres and Tickle, 1980):

Additionally, sensory integration theory is supported by three postulates (Botts, et al., 2008).

1. Learning is contingent on the ability of the student to receive sensory information, process the information, and integrate the information into a plan and organized behavior.

2. If the student has a deficit in processing and integrating sensory input, there will be a deficit in planning and producing behaviors.

3. Providing students with opportunities for sensory experiences enhances the ability of the central nervous system to process and integrate sensory information.

Specific principles and characteristics for delivering interventions using a sensory integration approach designed to assist students with autism are provided in the literature as requirements for effective sensory integration intervention techniques (Parham, et al., 2007):

- qualified professionals are involved;

- intervention is family-centered with appropriate assessment procedures;

- activities are rich in sensation including visual and auditory sensations;

- the intervention environment is safe and includes equipment that is free from injury;

- activities promote appropriate challenges;

- the intervention environment involves the whole body, moving and interacting with people and things in the three-dimensional space;

- the sensory environment intervention promotes intrinsic motivation and drive to interact through play; and

- the activities are their own reward with activities altered to meet the abilities of the student.

The concept of a multisensory environment (MSE) originated in Holland in the 1970s. These environments were originally named "Snoezelen," a contraction of two Dutch words: "snuffelen" meaning "to discover or explore," and "doezelen," referring to "a relaxed state" (Hulsegge and Verheul, 1987). The term emphasizes the idea of offering a soothing, nonthreatening, relaxing environment that promotes a general feeling of restoration and refreshment. Both children and adults can benefit from these environments, often with the close support of caregivers or partners, in pleasurable exploratory experiences that provide gentle stimulations without pressure to perform, and can be enjoyed full-body. Interventions in traditional MSEs have two primary goals: (1) stimulating the vestibular, proprioceptive, and tactile sensory systems to improve the individual's ability to process and integrate sensory stimuli in a correct way; and (2) promoting a general feeling of restoration and relaxation. In the past they offered no or limited capability for children to interact with physical artifacts and the whole space and to receive stimuli in a cause-effect mode (which is fundamental to develop cognitive skills).

Researchers exploring the educational and therapeutic effectiveness of Snoezelen methods have demonstrated improvements of the adaptive behavior and mitigation of stereotypical movements during the sessions in the MSE (Iarocci and McDonald, 2006; Stephenson and Carter, 2011). Collier and Truman (2008) explored the use of multi-sensory activity for individuals with multiple disabilities as a leisure resource and found that multi-sensory environments, when used as a companion to routine daily activities, enhanced sensory awareness and assisted with many behavioral challenges (e.g., aggression, agitation, wandering, poor coordination, and other difficulties). This combination of effects can enhance individual engagement and participation and reduce environmental barriers. Chan and colleagues (2007) reported mixed results from their research efforts to evaluate the clinical effectiveness of multi-sensory therapy on individuals with learning disabilities within a hospital setting. Their findings suggest that multi-sensory environments have a leisure resource effect of promoting psychological well-being rather than a therapy for reducing problem behaviors, and that reliability, predictability, relaxation, and freedom from demands, rather than sensory input, may be key contributors of multi-sensory therapy.

Traditional MSEs have not been used to facilitate learning scenarios with sequences or combinations of stimuli from multiple sources. However, the advances of network and sensor-based technologies commonly referred to as the Internet of Things opens possibilities for a new generation of "smart" environments. In these combined digital and virtual MSEs, various sensory affordances can be digitally connected, controllable, and interactive, and new intervention methods can be explored. Literature supporting the use of these latest multi-sensory interventions for people who identify as autistic points to some areas of potential positive outcomes.

SensoryPaint (Ringland et al., 2014) is a multimodal system that allows users to paint on a large display using physical objects, body-based interactions, and interactive audio. Researchers evaluated the impact of SensoryPaint through two user studies: a lab-based study of 15 children with neurodevelopmental disorders in which they used the system for up to 1 hour, and a deployment study with 4 children with autism, during which the system was integrated into existing daily sensory therapy sessions. Results demonstrated that a multimodal large display, using whole-body interactions combined with tangible interactions and interactive audio feedback, balances children's attention between their own bodies and sensory stimuli, augments existing therapies, and promotes socialization. These results offer implications for the design of other ubiquitous computing (ubicomp) systems for children with neurodevelopmental disorders and for their integration into therapeutic interventions.

Parés et al. (2004, 2005b) developed MEDIATE, a multisensory environment created to facilitate interaction and play in children with severe autism or those who are nonverbal. The immersive environment uses large projection screens with audio and a number of environmental motion sensors.

Figure 6.10: Sensory Paint: system showing the user's shadow on top (left). A user playing with the coloring book (center, right).

Individuals in the space can move about freely, and the large display and audio responds to their body movements based on changing pitch, tone, and speed. The goal was set by authors as for the children with autism to have fun and have the chance to play, explore and be creative in a predictable, controllable, and safe space where psychologists could better understand autism and the possible underlying communication mechanisms. Eleven children have participated in MEDIATE, each having three sessions in the environment. The time spent in the environment varied from 5–35 minutes. None of the children experienced discomfort in the environment, and only one of the sessions had to be stopped because of the child's overexcitement.

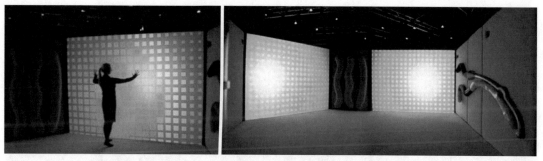

Figure 6.11: **MEDIATE** multisensory environment (Parés et al., 2004).

Parès et al. also developed "Lands of Fog" (Mora-Guiard et al., 2016), a full-body interaction experience in which an autistic child played together with a non-autistic peer. The system was aimed towards fostering social interaction behaviors and collaboration, and the researchers undertook user trials with 34 children on the autism spectrum through which the system has proven to be a useful tool to foster social interaction. Similarly, Crowell et al. (2018) developed GenPlay, a virtual environment system, which can rapidly transform a standard classroom into an interactive play space. Researchers observed social interaction consistently when players were focused on the exploration of the generative graphics. Lastly, Cibrian et al. (2017) used a fabric-based exploration of lights and colors to explore music and sounds with children with autism.

Figure 6.12: **Left:** Lands of Fog (Mora-Guiard et al., 2016). **Center:** GenPlay (Crowell, et al., 2018). **Left:** BendableSound (Cibrian et al., 2017)

Ahù is a nomadic multisensory solution for children's playful learning in schools (Gelsomini, et al. 2018). Ahù has a combination of features that address playful learning and inclusion (for children with and without autism): It has a totem shape; supports multimedia stimuli by communicating through voice, lights and multimedia contents projected on two different fields on the floor; and allows cooperative and competitive games. The system is nomadic, so it can be moved around, and can be controlled with easy input methods. An exploratory study investigated Ahù's potential to promote engagement, collaboration, socialization, and learning in classrooms.

Figure 6.13: Ahù projecting two play fields on the floor (Gelsomini et al., 2018)

Brown et al. (2016, Brown and Gemeinboeck, 2018) designed an interactive multi-sensory space called the "Responsive Dome Environment" (RDE) to elicit social communication between children and carers using participatory design approaches. RDE is a dome-shaped tent structure with a floor diameter of three meters and constrcted from a translucent acrylic texture. The material was chosen to diffuse colored lights surrounding the structure along with a quadriphonic sound system.

Figure 6.14: Responsive Dome Environment (Brown and Gemeinboeck, 2018).

Finally, a team of researchers (Garzotto and Gelsomini, 2016) co-designed with autism and education experts and developed a Magic Room called Magika, a playful supervised smart space for children with special needs, including autism. Results of an exploratory study involving 8 caregivers and 19 children who used the Magic Room across four months are encouraging in the cognitive (higher attention and memory recall), relational (more communicative occurrences), emotional (higher emotion expression), physical (stronger attitude in movement), and behavioral (reduction of problematic behaviors). The authors (Garzotto et al., 2016b) also explored ways of using brain signals, acquired through a wearable EEG headset, to have the space becoming responsive to a particular child's levels of relaxation and attention, automatically adapting the stimuli to the child's state, and allowing therapists to automatically collect a child's data for diagnosis purposes or to fine tune their therapy.

6.3 CLASSIFICATION APPLIED TO VIRTUAL REALITY, AUGMENTED REALITY, AND MULTISENSORY ENVIRONMENTS

In Figure 2.1, we tagged five representative articles as using virtual or augmented reality as a technology platform. These included Sam—a virtual peer (Tartaro and Cassell, 2008); virtual residential street, school playground, and school corridor scenes (Wallace et al., 2010); Social Mirror (Hong et al., 2012); Magika (Garzotto and Gelsomini, 2018); and Ahù (Gelsomini et al., 2018). Here we describe how each of these fit into the classification scheme defined in Chapter 2, as well as discuss overall trends we observed for technologies making use of virtual and augmented reality platforms.

> **Sam—a virtual peer** (Tartaro and Cassell, 2008): Sam—a virtual peer was classified as virtual and augmented reality because it includes augmented reality and a virtual avatar. Because the child talks to the peer and uses interactions with objects, its input modalities can be classified as *audio* and *physical* and output modalities as *audio* and *visual*. We classified it as targeting both *social/emotional* and *language/communication* skills given its focus on evaluating contingent discourse with Sam that involved turn taking, listening, and responding in a collaborative narrative. The goal of the system is *scientific assessment,* and the target end users are *persons with autism, clinician/therapist,* and *researchers.* The setting for the work was a *research lab* and the work was published in the proceedings of the International Conference of the Learning Sciences, an *educational* venue. Empirical evaluation included comparing measures of Theory of Mind and contingent utterances between a small group of autistic children and a small group of non-autistic children as they interacted with a human peer vs. virtual peer in a counterbalanced design. Therefore, we categorized it as *experimental.* Sam was evaluated at the early stage of a *design concept,* as it used Wizard of Oz prototyping techniques in its evaluation to simulate different experiences.

Blue Room (Wallace et al., 2010): This work was classified as virtual and augmented reality because it includes virtual environments and virtual avatars using primarily *spatial* input and *audio* and *visual* output. We classified it as targeting *social/emotional*, with the goal of *scientific assessment* given its focus on evaluating sense of presence in virtual scenes and social attractiveness of a virtual avatar through self-report questionnaires. Target end users are *persons with autism* and *researchers*. The study was performed in a *research lab*, published in an *autism-specific* journal, and was *descriptive* in nature. The system was designed with autistic individuals as *testers* and is currently at the stage of a *functional prototype*.

Social Mirror (Hong et al., 2012): The Social Mirror was classified as virtual and augmented reality because it includes an internet-based application designed for access via a computer-based Web browser and augmented reality (i.e., full-length mirror displays embedded in the natural environment that embodies the social networking tool). It uses *spatial* and *physical* inputs and *audio* and *visual* outputs. We classified it as targeting *social/emotional* and *life/vocational* skills with the goal of enhancing *intervention/education* by enabling users to gather social advice on their personal appearance and hygiene. Target end users are *persons with autism, family/caregiver,* and *peer*, and the intended setting for the system is the *home* environment. The work was published in *Computer Supported Cooperative Work and Social Computing* (CSCW), a *computing* venue. The empirical support included focus groups and is thus *descriptive*. The system is currently at the stage of a *functional prototype* and involved autistic individuals as *users* and *testers* in its design.

Magika (Garzotto and Gelsomini 2018): Magika was classified as virtual and augmented reality because it makes use of object augmented virtual environments. It primarily uses *physical* and *spatial* input and engages users in a fully sensory environment with *audio, physical, tactile,* and *other* output. There are a number of domains that Magika targets as an *educational intervention* and *scientific assessment*, including *behavior, cognition, life/vocational, play, sensory/physiological,* and *social/emotional* skills. The target end users for Magika include the *individuals with autism,* their *peers,* and *professional* caregivers in a variety of settings, including *clinics, research labs,* and *schools*. Autistic individuals were engaged as users and testers in the design process, and the empirical support is primarily *descriptive* using a *fully functional prototype*. Magika was published in across multiple papers, all of which are in *computing* oriented publication venues.

Ahù (Gelsomini et al., 2018): Ahù is a nomadic multisensory solution for children's playful learning in schools. Its primary input modality is *spatial* as it involved children moving around the room, and the output modalities engage all the senses with *audio, visual, tactile*, and *other*. The target domains for Ahù include *behavior, play, sensory/ physiological*, and *social/emotional*, and its primary purpose is as an *educational intervention* in *school* settings with *persons with autism* and their *peers* being the primary users. Autistic individuals were involved in the design of Ahù as *users* and *testers*. Ahù was published at the ACM Conference on Interactive Surfaces and Spaces, and thus its publication venue is *computing*, and it was evaluated with a *description* study of a *fully functional prototype*.

While domain variety was observed across the literature reviewed in this section, the majority of work in virtual and augmented reality tended to focus on social/emotional and life/vocational skills. The purpose for most deployments was scientific assessment of autism in research labs for researchers to evaluate characteristics of autism during virtual peer interaction and exploration/ navigation of virtual environments. Publication venues were rather evenly split between computing and autism-specific venues, with most empirical support at the descriptive level. Nearly all systems reviewed are functional prototypes.

6.4   FUTURE DIRECTIONS

Research focused on virtual, augmented, mixed, and multi-sensory environments is exploding globally. In the years between the first and second edition of this book, dozens of new papers were published describing novel systems and approaches, empirical tests of efficacy, and ethical and moral challenges related to the use of these platforms. As this area matures, there will no doubt be additional challenges and opportunities. However, as detailed in a number of recent reviews (e.g., Bellani, et al., 2011; Parsons and Cobb, 2011; Parsons, et al., 2013; Rajendran, 2013; Wang and Reid, 2011), we see four particular issues demanding additional research now.

First, many studies do not include well-established results that provide reliable data. Most studies assessed the potential use of VR/AR for training autistic people and reported positive results. However, most main research to date included case studies with important observations and only preliminary results. Lack of comparison and controlled studies makes it difficult to draw conclusions upon the preferences of this special audience group. Many studies consisted of developing a system that the researchers thought would best serve the needs of individuals with autism and looking into its effects in terms of the main goal on a small group of users in a research context. More comparison studies designed towards the investigation of VR/AR/MR properties, tested by a larger number of participants in other contexts, would help next-generation reality-based training applications for autistic people in providing more comfortable, adaptable, and effective experiences. More controlled studies are also needed to provide convincing results to pass regulatory approval for insurance reimbursement in the U.S. healthcare context. Such studies appear to be relatively few in number due to the lack of standardization of VR systems (hardware and software) and the protocols of the different operating groups. Possible collaborations with service providers might help in finding more participants for future studies. Human-subjects research in this space is inherently difficult, requiring creativity in study design, but meta-reviews and additional larger-scale efforts should be considered. Similarly, because there are so many frameworks for evaluating sensory integration, researchers must look for elements of commonality and difference in both the qualitative and quantitative assessments of these environments to better support autistic individuals, parents, teachers, co-workers, therapists, and so on in the development of environments that will predictably enable any particular individual to succeed.

Second, few studies explicitly and systematically evaluate whether newly acquired skills in virtual and augmented reality transfer to other environments. Assessments across virtual and natural environments are needed to substantiate the claim that skills transfer between settings. If the researchers could identify the factors that improve the rate of generalization (such as increased/decreased visual fidelity or more/less number of virtual environments), future VR applications can apply these principles, which would result in better quality training. Moreover, it would be important to find out effects of several properties on training individuals with autism with these systems: effects of different information presentation methods, different levels of visual fidelity (especially low and high since most of the previous studies utilized medium levels), output modalities, different levels of dynamism and crowdedness of the virtual environments could be explored. Effects of first- and third-person view, presence of virtual avatar, effects of avatar's appearance on users and effects of incorporating personal attributes of the user into the virtual world (such as clothing, hair color and model, and general physical appearance) would also give insight into better virtual reality experiences.

Third, many of the virtual and augmented systems employed in studies to date are extremely expensive and require computer science experts to be developed and even support their use. Moreover, most of the previous studies explored the use of computer-based VR systems for training applications catering to neurodiverse individuals. Among the studies that included immersive systems, many used old-generation head-mounted displays incontrast with today's technology (e.g., Google Cardboard). Hence, more studies exploring ways of utilizing new-generation, highly immersive headsets for training individuals with autism might provide valuable insights. In the case of MSEs, researchers, clinicians, and others wishing to make use of sensory-based interventions will have to consider lowering the high costs of multisensory equipment and invest their time on training people to use the rooms or virtual equipment as it is developed.

More affordable and intuitive systems are needed to enable wide-scale deployment, evaluation, and adoption. We are hopeful that this will be possible in the near future, as new, affordable technologies become available. This includes new depth sensors—three-dimensional cameras used primarily for gaming such as Intel Real Sense and Microsoft Kinect; new head-mounted displays and viewers such as Google Glass—a lightweight augmented reality headset that super-imposes virtual components on a real-world view; and HoloLens—a self-contained, holographic computer mounted on a headset equipped with multiple sensors, advanced optics, and a custom holographic processing unit. Likewise, newer technologies are becoming less bulky and cumbersome, which will likely improve acceptance of individuals with autism who have sensory issues. However, the human support required to use these tools will likely remain as essential as it is today. Specialists might be employed to develop a configuration of an environment suitable for particular individuals, much like human factors and ergonomics specialists are employed in many workplaces today. Configurable environments can enable the "loading" of an environment for each person as well as the adaptation of that environment in response to changing symptoms and context.

Fourth, although sensory-based treatments, including MSE, have been used for nearly 40 years in a wide variety of contexts, the addition of new technological supports may disrupt a proven technique. Traditional MSEs have been shown to support self-regulation specifically, and in some cases to support sensory processing more broadly. Development of technology-enabled MSEs increases the possibility of scalable and sustainable sensory-based interventions. Traditional MSEs require a dedicated physical space, often within a clinic or school, and can get overscheduled quickly. These environments are expensive to build or install, difficult to configure and reconfigure, and often adhere rigidly to a specific therapeutic intervention plan. On the other hand, technology-supported MSEs have the potential to be moved to home environments, turned on and off quickly, configured—even remotely—to support a variety of interventions and therapeutic approaches, and support a variety of goals.

A well-established design literature based on the perspective of users with autism, as well as others impacted by these systems, would serve as a valuable repository for future applications, helping produce a more holistic picture of needs and experiences. While virtuality is an exciting and promising new area, especially for training and targeted intervention, more research is needed to fully understand this nascent field.

CHAPTER 7

Sensor-Based and Wearable

In this chapter, we describe sensor-based and wearable technologies that have been used in relation to autism. Based on our classification scheme's definition, sensor-based and wearable technologies *include the use of sensors* (e.g., *cameras, microphones, peripheral physiological,* and *accelerometers*), *both in the environment and on the body, or computer vision to collect data or provide input.*

7.1 OVERVIEW

The technologies reviewed in this chapter are commonly referred to as "ubiquitous computers" (Weiser, 1991), "wearable computers"(Starner, 2001, 2002), the Internet of Things, and collectively as "telemetrics" (Goodwin et al., 2008). The commonality among them is their discreet size or form factor, capability of measuring data wirelessly, and ability to produce and transmit synchronized, time-stamped datasets to remote locations for viewing and analysis.

The motivation behind the use of sensors in ubiquitous computing is to "weave themselves into the fabric of everyday life until they are indistinguishable from it" (Weiser, 1991, p. 94). They involve embedding sensing technologies in the environment, including objects within them, and constitute "living laboratories" and "smart rooms" capable of wirelessly monitoring surroundings and inhabitant behavior (Abowd et al., 2000; Intille et al., 2005; Kaye et al., 2011). The instrumented spaces and objects within them often include sensors to record interior conditions (temperature, humidity, light, etc.), person-object interactions (e.g., radio frequency identification (RFID) attached to common items), and human behaviors (video cameras, microphones, motion sensors, etc.).

"Wearables" are on-body perception systems sewn into articles of clothing or embedded into accessories such as shoes, gloves, glasses, and jewelry (e.g., Healey, 2000). For example, small, wireless peripheral physiological sensors have been developed to unobtrusively record cardiovascular, respiratory, and electrodermal activity in freely moving people (Wilhem and Grossman, 2010). Miniature actigraphs and accelerometers capable of objectively quantifying posture and physical activities have been embedded in wristbands, bracelets, and belts (Bao and Intille, 2004). Wearable audio-capture technologies that utilize small, unobtrusive microphones to record sounds created by the user (e.g., speech, gestures) and ambient auditory events in the environment (Mehl et al., 2001; Stager et al., 2003) have also been developed. And discreet cameras integrated into eyeglasses to determine where a user is looking and what she or he is seeing are also becoming increasingly available commercial products used in scientific research (e.g., Tobii,[18] Positive Science[19]).

7.2   SENSOR-BASED AND WEARABLE TECHNOLOGIES AND AUTISM

Ubiquitous and wearable sensors support the following three key advances in autism research and intervention efforts: ecological validity, repeated assessment, and reduced reactivity to measurement.

Ecological validity refers to the relation between assessments of behavior in experimental contexts and behavior as it is produced in the real world (Brunswik, 1947; Schmuckler, 2001). Often used synonymously with *generalizability*, ecological validity connotes *representativeness* or *naturalness*. The overwhelming majority of autism-related research is conducted in laboratory, hospital, and clinical settings. However, the most ecologically valid behavior assessment strategies are those that make observations in the real world where behavior, and all of its structural and functional relationships, occur naturally. For those who identify as autistic, these include home, school, work, and other community settings.

Conducting repeated assessments in autism-related research efforts is critical given heterogeneity in symptom presentation and varied developmental trajectories observed in the population (Gotham, Pickles, and Lord, 2012). Both of these factors constitute important individual differences that can be obscured by averaging responses across people and relying on few measurement types and points in time.

Reactivity refers to the phenomenon of measurement processes producing change in what is measured (Campbell and Stanley, 1996). Reactivity is an important factor when evaluating the experience of autism because most standardized assessments includes foreign and invasive procedures (e.g., fMRI, EEG, aptitude testing) conducted in unfamiliar settings (laboratory, hospital, clinical settings) with unfamiliar people (trained test administrators) that require enormous amounts of novelty and self-regulation with which to comply. This not only threatens internal, external, and construct validity, but can also create selection biases wherein only a small subset of autistic people might participate in research studies (the consequence of which is little representation from those with autism who may have more barriers to participation or may be more susceptible to issues of novelty; Stedman et al., 2019). A strategy for reducing the effects of behavioral reactivity is to use observation procedures that are passive (i.e., collect data without conscious input from the person being observed), involve little or no alteration of environmental stimuli, and minimize evaluation apprehension.

---

[18]   http://www.tobii.com/
[19]   http://www.positivescience.com/

As demonstrated below, a variety of ubiquitous and wearable sensors have been usefully employed with individuals on the autism spectrum (including those with a range of abilities), including telemetric video, audio, physiological, and physical activity sensors.

7.2.1   VIDEO ASSESSMENTS

As exemplified in this chapter, ubiquitous and wearable video sensors have been used to assess diagnostic and developmental status, monitor repetitive behaviors, capture salient life experiences, and teach social-emotional abilities to individuals on the autism spectrum.

When asked about initial concerns regarding their child with autism, at least 30–50% of parents recalled abnormalities dating back to the first year (Gillberg et al., 1990; Hoshino et al., 1987; Volkmar et al., 1985). Similarly, studies of early home videos reveal behaviors indicative of autism in children later diagnosed compared to those of typically developing children (Baranek, 1999; Mars et al., 1998; Osterling and Dawson, 1994). According to both information sources, children with autism in the first year of life are distinguished by a failure to orient to name, decreased orienting to faces, reduced social interaction, absence of social smiling, lack of spontaneous imitation, lack of facial expressions, lack of pointing/showing, and abnormal muscle tone, posture, and movement patterns.

Although parents' retrospective reports and home video analyses clearly point to atypicality in an autistic child's early development, this body of research is potentially limited by a host of methodological problems (for a more detailed review, see Zwaigenbaum et al., 2007). For instance, a parent's incidental observations regarding subtle social and communicative differences may be limited compared to systematic assessments by trained clinicians. Parents' tendency to use compensatory strategies to elicit their child's best behaviors (with or without awareness) may also affect their behavioral descriptions. Retrospective parental reports can also suffer from distortions of recall, especially when parents are asked to remember behaviors that occurred many years earlier. They can also include significant inaccuracies with respect to the description and perceived timing of early behavioral signs associated with autism. Finally, environmental manipulations and systematic presses for specific behaviors cannot be controlled for in retrospective studies.

Home video analysis has significant strengths over retrospective parental reports as it allows the observation of behaviors as they occur in familiar and natural settings. They also enable more objective ratings of behavior by trained observers who are less biased. However, this methodology also has limitations. The primary shortcoming is that parents typically record videotapes to preserve family memories rather than document their child's behavior systematically over time. As a result, footage from different families varies as a function of length of time the child is visible, activities recorded, and quality of recordings. Moreover, if children do not behave as expected or desired, parents may re-record taped segments until they obtain more favorable responses. Observations from home videos also vary considerably between children and depend on particular contexts selected for taping. Another potential problem relates to sampling contexts of home videotapes in so much as they may not provide sufficient opportunities for social communicative behaviors to be adequately assessed.

To overcome some of the shortcomings of home video recording, ubiquitous video systems are being deployed prospectively inhome settings to more fully capture, quantify, and communicate early behavioral manifestations of autism. For example, Vosoughi and colleagues (2012) created the Speechome Recorder, a portable version of the audio/video recording technology developed for the Human Speechome Project (Roy et al., 2006) (see Figure 7.1). The Speechome Recorder is a lamp-like form factor containing a dual camera system—one overhead camera facing down and the other camera facing horizontally at the height of a young child. Both cameras use 185-degree angle-of-view lenses able to record at 15 frames per second at a resolution of 960 × 960 pixels, enabling determination of interactions with surrounding people and objects throughout a room. The system also includes a boundary layer microphone that uses the surface in which it is embedded as a pickup. This allows a microphone placed in the head of the recorder to pick up speech in any corner of the room. All data captured by the Speechome Recorder can be stored locally on device and/or transmitted securely over ethernet. Incipient results from this investigation were published in Chin et al. (2018) and demonstrate that densely collected audio-video behavioral samples in the home reveal subtle but important language production differences in a child with autism that would be missed if only few traditional clinic-based measures of acquisition were used.

PHOTO: (C) PHILIP DECAMP AND DEB ROY 2012 FROM THE HUMAN FACE OF BIG DATA

Figure 7.1: Video feeds from the Human Speechhome Project (Roy et al., 2006).

Wearable point-of-view cameras have been used to automate measures of eye gaze (Edmunds et al., 2017), eye contact (Jones et al., 2017; Ajodan et al., 2019), and dyadic social interactions (Rehg et al., 2014) in autistic people in both clinic and naturalistic (e.g., home, workplace, school) settings. Computer vision researchers have also attempted to detect and characterize stereotypical motor movements in individuals with autism by means of automated analysis of captured video (Goncalves et al., 2012; Ciptadi et al., 2014; Jazouli et al., 2016).

Difficulty communicating and engaging in real-time social interactions is another core characteristic often described as part of an autism diagnosis, including those who have verbal abilities and normal to above average intelligence (Klin et al., 2005). As a result, social interactions can be complex, confusing, and tiring for many autistic people, a point often made by autistic self-advocates as well. Non-autistic people should do more work to support their end of this communication and social interaction. The result of the currently exhausting social interactions is that it can be difficult for autistic people to establish peer networks and work and learn in traditional educational and workplace environments. Schools and workplaces simply must be more inclusive and accommodating. At the same time, to support those autistic people who wish to better understand and interact with their non-autistic community members, Madsen et al. (2009b) created iSET (interactive Social-Emotional Toolkit) to support autistic people better in everyday conversations. Through a technology-augmented game, participants were assigned the goal of capturing facial expressions from their teachers and peers using tablet computers running real-time facial expression inference algorithms. The system also included an offline component where an individual with autism and their teacher could review previously recorded video together and learn about ecologically valid facial expressions at their own pace.

Using a somewhat similar technology-enabled approach, Hayes and colleagues (2010) and Marcu and colleagues (2012) deployed Microsoft SenseCams (Hodges et al., 2006) to periodically capture the views of the environment from the perspective of autistic children. These cameras automatically capture images based on changes in onboard sensors (light, temperature, accelerometer) throughout the day (see Figure 7.2). The purpose of these deployments was to enable minimally verbal autistic children to capture and share life experiences with caregivers and teachers in home, school, and community settings, and to populate more personalized and situated picture-based communication systems.

Figure 7.2: SenseCam wearable camera used with children on the autism spectrum (Hayes et al., 2010).

7.2.2   AUDIO ASSESSMENTS

Communicative style, including speech and language, is one of the primary differences seen people with autism when compared with their non-autistic peers. In some cases, these differences are more marked and may make it difficult or impossible for others to understand the communication, but in many cases, these differences can be overcome by work on both sides of the communication. For instance, autistic children are found to communicate less frequently when young; develop language later; produce atypical patterns of sound or vocalizations and repetitive speech; and have conversational and narrative skills that their non-autistic communications partners struggle to understand (Tager-Flusberg et al., 2011). Ubiquitous and wearable microphones have been used to detect prodromal speech features in infants at high risk for autism (Swanson et al., 2018); discriminate between autistic, delayed, and typically developing children's speech using computational markers of repetitive speech (van Santen et al., 2013); prosody (Chaspari, et al., 2012; van Santen et al., 2010); vocalization frequency (Oller et al., 2010); and vocalization composition (Xu et al., 2009). Kinsella, Chow, and Kushki (2017) have also demonstrated that a Google Glass-based application called "Holli" can increase social skills in individuals with autism by audio recording conversation, transcribing it into text in real time, and providing social prompts visually to the person wearing Google Glass. Finally, Boyd et al. (2016) demonstrated that both autistic and non-autistic people can collectively regulate volume working together with the help of a wearable application on Google Glass.

7.2.3   PHYSIOLOGICAL ASSESSMENTS

Communication and socialization difficulties, emotion dysregulation, sensory processing challenges, differences in executive function, and behavioral rigidity common in many people with autism can make them exceedingly vulnerable to stressors and limit their available coping strategies (Baron et al., 2006). Ineffective coping to stressors can lead to anxiety, and research suggests comorbid anxiety is present in 33–84% of individuals with autism sampled (van Steensel et al., 2011; White et al., 2009). However, most of this research is based on parental report measures since many individuals with autism either lack communication abilities altogether or, for those with language, have difficulties identifying and describing their feelings through self-report (Hill et al., 2004; Costa, Steffgen, and Samson, 2017). This approach leaves a big hole in the available research data to handle the possibility that caregivers and clinicians have misunderstood autistic people and limits the generalizability beyond a certain age, typically early adulthood. It can also be difficult for observers to infer internal arousal states in autistic people given they frequently exhibit reduced behavioral and affective expressions (Tordjman et al., 2009). Researchers in recent years have attempted to overcome unreliable self-reports and reliance on behavioral and affective observations by wirelessly recording peripheral physiological activity to specific stressors in freely moving individuals on the autism spectrum (Goodwin et al., 2006; Groden et al., 2005; Kushki et al., 2013). Moreover, researchers have employed telemetric physiological monitors to evaluate diagnostic symptoms (Prince et al., 2017), sensory responses (Woodard et al., 2012), social stressors (Levine et al., 2014), and affective responses (Liu et al., 2008b; Sarabadani et al., 2018) in individuals with autism.

In more recent work, researchers have begun to use wearable physiological sensors to better understand and predict challenging behaviors exhibited by autistic people. For example, using three-axis accelerometry and a telemetric cardiovascular monitor, Heathers and colleagues (2019) demonstrated that both repetitive body rocking and hand flapping associate with a strikingly similar cardiovascular pattern of acceleration and deceleration unrelated to physical demands associated with the movements themselves, suggesting a putative mechanism for homeostatic and somatosensory functions of these behaviors. In a somewhat similar line of research, Goodwin and colleagues (2019) investigated whether preceding physiological and motion data measured by a wrist-worn biosensor can predict aggression to others by autistic youth in a hospital inpatient setting. Developing prediction models based on ridge-regularized logistic regression, their results suggest that aggression to others can be predicted 1 min before it occurs using 3 min of prior biosensor data, with an average area under the curve of 0.71 for a global model and 0.84 for person-dependent models. These findings lay the groundwork for the future development of antecedent behavior analysis and just-in-time adaptive intervention (Nahum-Shani et al., 2018) mobile health systems (mHealth) (Kumar et al., 2013) to prevent or mitigate the emergence, occurrence, and impact of aggression in autism.

7.2.4   PHYSICAL ACTIVITY ASSESSMENTS

Accelerometry offers a practical and low-cost method of objectively monitoring the ways in which free-living humans move and manipulate objects. An accelerometer is an electromechanical sensor that measures static (constant force of gravity) and dynamic (moving or vibrating) acceleration forces. An active area of research with accelerometers is the measurement of physical activity in individuals on the autism spectrum, particularly those who engage in challenging behaviors.

Autistic people and their allies are commonly concerned about meltdowns and other physically adaptive but culturally unacceptable physical behaviors. Exhibiting such behaviors tends to have major impacts on their quality of life and that of their caregivers and can seriously affect the ability to reside in and benefit from less restrictive environments. Such behaviors can include aggression toward others, property destruction, self-injury (head hitting, biting, etc.), stereotypical motor movements (hand flapping, body rocking, etc.), and elopement (i.e., abruptly running away) (Cohen et al., 2011).

Standardized parent or teacher-report checklists (e.g., Rojahn et al., 2001), direct observation (Foster and Cone, 1986), and video-based methods constitute traditional ways of recording challenging behaviors in children with autism. However, they all have drawbacks, and most cannot be used to support autistic adults in their own efforts towards self-regulation. While filling out checklists can be useful and efficient, they fail to capture intra-individual variation in the form, amount, intensity, and duration of challenging behaviors over time (Pyles et al., 1997). Direct observation—which involves watching and recording a sequence of behavior in real time—can be unreliable due to the speed with which challenging behaviors can occur, difficulty determining when a behavior has started and ended, and ability to estimate behavioral quantities over a finite period of time (Sprague and Newell, 1996). Video-based methods—which involve video capture of behavior and offline coding and analysis— is more reliable than checklists and direct observation given the ability to review videos repeatedly and slow playback speeds. However, they are tedious and time consuming, making them expensive and impractical for most researchers to use (Matson and Nebel-Schwalm, 2007) and invasive towards the autistic children being recorded.

In an effort to overcome these methodological problems, researchers have begun exploring the use of wireless accelerometers and pattern-recognition algorithms to provide automated measures of stereotypical motor movements in individuals with autism that may be more objective, detailed, and precise than rating scales and direct observation, and more time and cost efficient than video-based methods. For instance, in a series of studies, Goodwin and colleagues (Albinali et al., 2009, 2012; Goodwin et al., 2011; Großekathöfer et al., 2017) demonstrated that stereotypical motor movements can be automatically detected and characterized with up to 90% accuracy in individuals with autism in both laboratory and classroom settings (see Figure 7.3). Automated recognition of other behaviors commonly produced by autistic people including aggression, self-injury, and forceful contact with objects in the environment (desks, walls, etc.), has also been conducted and yields promising results to assist in determination of eliciting stimuli and document response to intervention (Plotz et al., 2012). Finally, researchers have piloted the use of toys embedded with accelerometers and pattern recognition on resulting data to characterize play behavior in children with autism (Bondioli et al., 2019).

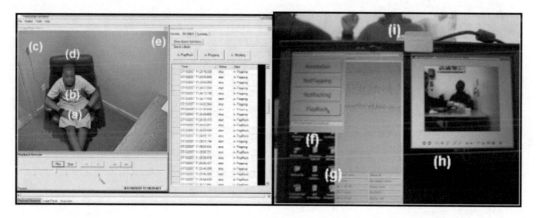

Figure 7.3: Body-worn accelerometers to sense stereotypical movements (Albinali et al., 2009).

7.3   CLASSIFICATION APPLIED TO WEARABLE AND SENSOR-BASED TECHNOLOGIES

In Figure 2.1, we tagged four representative articles as using sensor-based or wearable technologies as a technology platform. Here we briefly describe how each of these fit into the classification scheme defined in Chapter 2, as well as discuss overall trends we observe for technologies making use of sensor-based and wearable platforms with individuals who identify as being on the autism spectrum.

LifeShirt (Goodwin et al., 2006): We classified this work as *sensor-based and wearable* targeting the domain of *sensory/physiological/motor* as it employed an ambulatory cardiovascular measure (Lifeshirt) to assess physiological responses to potential stressors. This uses *physical* input (passively via sensors) and *visual* output (as data). The goal of the work is *scientific assessment* and *persons with autism* and *researchers* are the target end users. The study was conducted in a *school* setting, and results were published in *Focus on Autism and Other Developmental Disabilities*, an *autism-specific* journal. Empirical support was *experimental* comparing cardiovascular stress responses between children with autism and age- and sex-matched typically developing controls. The system used to collect cardiovascular responses was *publicly available* at the time of publication, but it is no longer manufactured commercially. It was developed with autistic individuals as *users*.

Three-axis accelerometry and telemetric cardiovascular monitor (Heathers et al., 2019): This research uses sensor-based and wearable computing, including *physical* input (passively, via worn sensors) and *visual* output (data visualizations) to address the domain of *restrictive/repetitive behaviors* as it sought to investigate the relationship between cardiovascular responding and stereotypical motor movements. The goal of the system is *scientific assessment* and targets *persons with autism* and *researchers* as end users. The work was carried out in both a *research lab* and *school* setting, and was published in *Biological Psychology*, a *social/behavioral science* venue. Empirical support was *correlational/quasi-experimental* and the system used to collect cardiovascular responses was *publicly available* at the time, but is no longer manufactured commercially. To the best of our knowledge, the devices involved autistic individuals as *users* in the design process.

Sensor-embedded toys (Bondioli et al., 2019): We classified this work as *sensor-based* as it incorporated toys instrumented with accelerometers, as well as a user interface to analyze play behavior. This uses *physical* input and *visual* output. The system addresses the domains of *social/emotional*, *play*, and *behavior* with the goal of *diagnosis/screening* and *scientific assessment*. Target end users of the system include *persons with autism*, *family/caregiver*, *professional*, and *researcher*. Settings for the work include the *home*, *clinical setting*, and *research lab*. Results from the study were published in *Springer Nature Switzerland*, a *social/behavioral science* venue. Empirical support for the work is *descriptive* and the system is currently a *functional prototype*, with individuals with autism included as *users* of this approach.

Wrist-Worn Biosensors (Goodwin et al., 2019): This work uses the E4—a commercially available device recording motion, electrodermal activity, and cardiovascular measures and pattern classification to predict aggression to others before it occurs. It has primarily *physical* input (passively, via sensors) and *visual* output (via data visualization). The goal of the system is *scientific assessment* and *intervention* for autistic people. The research targets *persons with autism*, as well as *researchers*, *caregivers*, and *professionals* as end users. The work was carried out in a naturalistic *clinical setting*. Results were published in *Autism Research*, an autism-specific venue. Empirical support was *correlational/quasi-experimental* and the system (E4) is *publicly available*. To our knowledge, E4 was designed with autistic individuals as *users*.

The work reviewed in this area is quite diverse, including both embedded and wearable video, audio, physiological, and physical activity sensors to facilitate diagnosis/screening, scientific assessment, and functional assessment in the areas of social/emotional, language/communication, and restrictive/repetitive behaviors. Target end users of these systems tended to focus on researchers with data collection carried out in research labs, though not exclusively. Publication venues for the work most often appeared in autism-specific and computing journals and proceedings. Empirical support varied but included a higher number of correlational/quasi-experimental and experimental findings than other interactive technology platforms covered in this book. Most systems reviewed are currently functional prototypes.

7.4   FUTURE DIRECTIONS

Based on the studies cited in this chapter and recent reviews of peer-reviewed published reports employing sensor-based and wearable technologies in autism (Cabibihan et al., 2017; Koumpouros and Kafazis, 2019), there are several potential future directions for the use of ubiquitous and wearable sensors in autism research and intervention. With notable exception of the computational speech and language studies cited, most of the examples in this chapter involve relatively small samples. Larger samples, ranging in age and abilities, would be useful to assess replication and generalizability of findings (i.e., a recent meta-analysis concluded that only 25% of participants with autism across 301 studies had an intellectual disability; Russell et al., 2019). Extending deployments of these sensors for longer periods of time in natural settings, especially with non-technical researchers who were not involved in developing the systems but have domain expertise in autism, would demonstrate system reliability, validity, and robustness. It would also be interesting to see if sensor fusion and analytic optimization could be brought to bear on sensor-based and wearable data to enable just-in-time feedback that could be used to assess, evaluate, and intervene with individuals on the autism spectrum in real time.

CHAPTER 8

Natural User Interfaces

In this chapter, we describe what has come to be termed "natural user interfaces" (NUIs) and review how they have been used with individuals on the autism spectrum. Although the word natural can be hotly contested given that most movements are not actually "natural" but learned, we use the standard phrasing of natural user interfaces in this scope of the review. By this definition, "natural" input encompasses a large variety of input techniques beyond the traditional desktop computer, such as the use of pen, gestures, speech, tangible computing, and eye-tracking technologies. Here, we define NUIs as involving the use of input devices and techniques beyond traditional mice and keyboards. Specific input techniques include penwriting, gestures, speech/voice, eyetracking, tangible computing, and smart objects. Natural user interfaces also involve interaction with a system rather than just providing passive input.

8.1   OVERVIEW

Hinckley (2002) defines an interface as natural if "the experience of using a system matches expectations, such that it is always clear to the user how to proceed, and that few steps (with a minimum of physical and cognitive effort) are required to complete common tasks." Although Hinckley does not name specific technologies, interactions that use gestures, speech, touch, and gaze are types of input that fall into this category. Abowd and Mynatt (2000) further describe the application area of natural interactions as it relates to third-generation ubiquitous computing technologies.

Although some would consider multi-touch surfaces such as iPads, Microsoft Surface, and more to be NUIs, we exclude them from this category because they are included in shared active surfaces discussed in Chapter 5. We also contrast NUIs with sensors or wearable devices discussed in Chapter 7, as these techniques tend to focus on sensing passive input rather than back-and-forth explicit interactions. Finally, we also exclude NUIs that are incorporated as part of a multi-sensor or A/R environment, which we cover in Chapter 6.

NUIs have seen success when used in the autism domain, likely because they enable a variety of input mechanisms specific to the needs of people who present with different sensory processing challenges. For example, if an autistic person has problems with their visual channel, they might be able to use speech-based interaction or physical gestures as input. However, there are some challenges in this space, as some types of channels may be problematic for different individuals. In the case of an individual with autism who exhibits stereotypical motor movements, such as body rocking or hand flapping, it may be difficult to ascertain communicative physical gestures. Other natural input types that do not use motion, such as eyetracking, might be more suitable for individuals who exhibit these behaviors, though may also present a challenge if individuals engage in too much movement. Another primary advantage of NUIs is they often involve a lower learning curve for people with cognitive impairment, and thus may be easier to use than a keyboard or more complex input device. They also have the advantage of being nearly ubiquitous, and thus may reduce stigma associated with having to use, carry, or wear a specialty device (Shinohara and Wobbrock, 2011). Finally, NUIs have become more prevalent over the past decade, which has increased their prevalence and lowered their cost, making them a more practical technology platform for individuals with autism.

In general, NUIs have the largest variety of inputs compared to other types of technologies we reviewed, enabling a multitude of applications. Voice, pen, and physical gestures can be used as alternative inputs to almost any existing technology, and some technology designers have used these types of interactions to make applications easier for people with a wide variety of disabilities, including autism, to use. New applications are also being developed to specifically take advantage of these types of NUIs for autism, such as games and therapy applications. In the subsequent section, we discuss specific examples that take advantage of these platforms.

8.2   NATURAL USER INTERFACE TECHNOLOGIES AND AUTISM

In this section, we describe specific technologies deployed with people who identify as on the autism spectrum that include pen and physical gesture-based interactions, tangible computing, voice input, speech recognition, and eye-tracking systems that are interactive.

8.2.1   PEN AND PAPER

Pen gestures have the advantage of blending seamlessly in environments in which they are used. Because individuals with autism are often reported to be enamored of technology, using more obvious devices like tablets and laptops may serve as a distraction in classrooms or therapy situations (Kientz and Abowd, 2008). Pens in particular are advantageous in this regard, since many practices within the autism community are still very much paper based (Marcu et al., 2013). One trend for NUI systems has been to include digital pens as an input mechanism for therapy sessions. For example, the Abaris application discussed in Chapter 4 (Kientz et al., 2005, 2006; Kientz, 2012) combines audio and video recording with digital pen and paper (using Anoto technology[20]) and speech recognition to capture data from therapy sessions to allow better recordkeeping and collaborative review (see Figure 8.1).

---

[20]   http://www.anoto.com

The Talking Paper project (Garzotto and Bordogna, 2010) uses paper in a different way, where paper-based objects, such as PECS cards, drawings, and pictures, are augmented and connected to an on-screen device that would play videos or interactive media as the paper was manipulated. Although Talking Paper was primarily designed with neurotypical children, it also has seen some success with autistic children in school settings.

Figure 8.1: Left: screenshot of Abaris video review screen. Center: digital pen and paper input form (Kientz et al., 2006, 2005; Kientz, 2012). Right: Talking Paper (Garzotto and Bordogna, 2010).

8.2.2   TOUCHLESS GESTURES

A number of projects detailed in shared active surfaces, robotics, virtual reality, and sensor and wearable chapters in this book have used gestures as a means of interaction for autistic users of these systems. Gestures include both natural movements like pointing and dragging, as well as learned gestures such as pinching, panning, or tapping. Gestures can be arbitrary or symbolic, such as those used in American Sign Language. Recent advances in the use of depth cameras and computer vision processing have enabled a number of touchless gesture-based applications, including the use of the whole body, to be more widely used. We refer to these as motion-based touchless gestures to distinguish them from other gesture-based interactions used with touchscreens, which we describe in Chapters 4 and 5.

Microsoft Kinect is the most prominent commercial example of a platform for supporting touchless gestures, and the release of a toolkit for developing applications on top of the Kinect platform has enabled a wide variety of new gesture-based interactions. These new applications have been designed to improve motor skills, body awareness, attention, concentration, and social behavior. Boutsika (2014) describes how the Kinect can be used in educational settings for autistic students. There is also evidence that gestures can improve learning outcomes for children (Broaders et al., 2007). There are several prominent examples of the use of Microsoft Kinect for individuals with autism. Commercially available games, such as Kinect Sports and Rabbids Alive and Kicking (Figure 8.2), have been used by autistic children to practice motor skills (Bartoli et al., 2013; Garzotto et al., 2014).

Figure 8.2: Kinect games evaluated by individuals with autism.

A project called S'Cool involved an in-depth exploration of the use of motion-based activities specifically for classroom use and included a nine-month observational study on how teachers and students used motion-based games for children and their peers in an autism-specific classroom setting (Bhattacharya et al., 2015). Their methods included prototyping new ideas (see Figure 8.3, left) and identified positive overall impacts of the technology for engaging with peers and show the feasibility of in-classroom gesture-based interactions. Other research has used simple webcam based Web games for similar educational activities (Li et al., 2012), while Garzotto and colleagues developed Pixel Balance (Figure 8.3, center and right) to support imitative capability, body schema awareness, and social skills (Garzotto et al., 2014).

Figure 8.3: S'Cool prototype app (left) and Pixel Balance (center and right).

8.2.3   TANGIBLE COMPUTING AND SMART OBJECTS

The greatest number of projects in the area of NUIs relating to autism has been in the area of tangible or tactile computing (the terms are often used interchangeably). Tangible computing is the use of physical objects that can be grasped and manipulated to interact with a virtual environment or system (Ishii and Ullmer, 1997). Tangible objects may include building blocks, small figures, plush dolls or animals, robots, toys, or other custom-made objects. The interaction may occur locally on the device itself, with a complementary display, or with an entire room environment. They typically combine both the physical and the digital, and the applications often use RFID tags and readers, computer vision, or other sensors to recognize when and how a user interacts with an object.

One study (Sitdhisanguan et al., 2012) compared the use of tangible user interfaces, touch-based interfaces (e.g., those described in Chapter 4 and 5), and desktop-based interfaces for computer-based training with individuals on the autism spectrum (e.g., those described in Chapter 3). Results suggested that tangible and touch-based interactions were easier to use, and that tangible systems were more effective teaching tools than standard desktop-based applications or non-computer-based systems. Another study showed that even a simple device that vibrates in a child's pocket was a promising approach in rewarding verbal initiations (Taylor and Levin, 1998). Others have done general explorations of different types of technology within the tangible space (Keay-Bright and Howarth, 2012), as well as explored the use of tactile feedback with a robot outfitted with special skin that could respond to and record a child's touch (Amirabdollahian et al., 2011).

Tactile computing can be especially useful for people who are more drawn to virtual worlds than to the physical world initially, a phenomenon commonly confirmed by people who identify as on the autism spectrum as we described in Chapter 6. Tactile, tangible, and multi-modal computing can be used to scaffold experiences to enhance engagement and interactions in the real world. Farr and colleagues (Farr et al., 2010a) describe tangibles as having the ability to "provide a safety net for encouraging social interaction as they allow for a broad range of interaction styles." Their work included an augmented knight's castle toolset that used RFID tags to activate sounds as the child plays with an object, including recorded audio in a child's voice (see Figure 8.4). They conducted a small experiment on this technology to determine whether configurability of the knight's castle toy set encouraged social interactions. They found it encouraged greater occurrence of orientation behaviors and more parallel and cooperative play than solitary play.

Figure 8.4: Augmented Knight's castle game (Farr et al., 2010b) that uses RFID tags and readers to identify when the player interacts with objects.

Topobo is another tangible user interface construction kit (see Figure 8.5, left) designed for autistic children, which is a programmable environment that is easy to learn (Farr et al., 2010a). They compared the use of Topobo and standard LEGO blocks in a small experiment with six children, all of whom had autism diagnoses. Their study indicates that the programmable Topobo set elicited higher levels of social interactions than just LEGO, as well as more parallel play rather than solitary play. Other technology approaches have also used building-block type toys with electronic enhancements to promote creativity, interaction, and teamwork (Drain et al., 2011).

Affective Social Quest, by Blocher and Picard (2002), used plush dolls combined with an on-screen video to help teach emotion recognition to autistic children (see Figure 8.5, right). In the application, the on-screen video displayed an emotion, and the child was tasked with finding a physical doll that matched the expression. When the correct doll was found, a wireless two-way communication protocol triggered rewards both on screen and via the doll itself, which vibrated and displayed an emotion similar to the one displayed on screen (e.g., giggling for a happy doll). The system was evaluated with six autistic children who used the tool across three days to determine its feasibility.

Figure 8.5: **Left:** Topobo (Farr et al., 2010a). **Right:** Affective Social Quest (Blocher and Picard, 2002).

One newer smart object includes Polipo (Tam et al., 2017), a tangible 3D-printed toy (Figure 8.6) that can be customized for therapy with autistic children. Designed with collaboration from therapists, Polipo can be used to teach fine motor skills and allow new sensory explorations. Other augmented object examples include "smart" chopsticks, EasyClap and EasyPick, that autistic children use to improve dexterity and fine motor control (Chia and Saakes, 2014); ThingsThatThink (Escobedo et al., 2014) which are tools that allows for augmenting of objects to promote interactivity and recording of certain behaviors; and MoSo (Bakker et al., 2011), augmented objects tied with sounds to explore different physical manipulations.

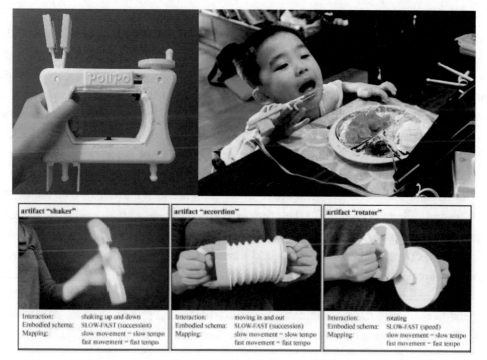

artifact "shaker"	artifact "accordion"	artifact "rotator"
Interaction: shaking up and down Embodied schema: SLOW-FAST (succession) Mapping: slow movement = slow tempo fast movement = fast tempo	Interaction: moving in and out Embodied schema: SLOW-FAST (succession) Mapping: slow movement = slow tempo fast movement = fast tempo	Interaction: rotating Embodied schema: SLOW-FAST (speed) Mapping: slow movement = slow tempo fast movement = fast tempo

Figure 8.6: **Top left: Polipo** (Tam et al., 2017). **Top right: EasyPick** (Chia and Saakes, 2014). **Bottom: MoSo** (Bakker et al., 2011).

In addition to using tangible computing for children already diagnosed with autism, there have been efforts to use sensors embedded in tangible objects, such as children's toys, to understand how young infants play with objects and to look for early warning signs of autism, such as repetitive play or self-stimulatory behaviors (Westeyn et al., 2012) (see Figure 8.7).

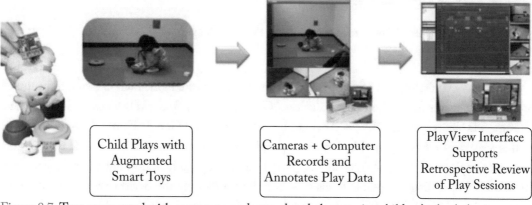

Figure 8.7: Toys augmented with sensors to understand and characterize children's play behaviors (Westeyn et al., 2012).

8.2.4   VOICE, SPEECH, AND AUDIO

Because autism often co-exists with limited speaking or difficulty with spoken language that is either persistent or contextual, there have been fewer applications that make use of speech recognition as a way of interacting with a computerized system. The use of audio and speech processing has largely focused on assessing speech in children or for recognizing speech prompts from adult caregivers. For example, a number of the applications discussed in Chapter 7 used speech as an assessment tool, such as measuring repetitive vocal behaviors. The Abaris system mentioned previously also used speech recognition, but focused on the speech of the therapists working with children for indexing video streams (Kientz et al., 2005, 2006).

One notable example of an interactive technology for use with individuals with autism is that of Hailpern and colleagues (2009a). In their system, children engage with an interactive system that visualizes their non-speech vocalizations. The idea is to encourage these vocalizations in non-verbal autistic children through animations, sound, and stimulating rewards. Hailpern and colleagues have deployed this system with numerous children on the autism spectrum in a lab setting and found they could be used to stimulate any vocalization with the Spoken Impact Project (Hailpern et al., 2009b) and to encourage multi-syllabic vocalizations with VocSyl (Hailpern et al., 2012) (see Figure 8.8). The design process for these systems is also described (Hailpern et al., 2012).

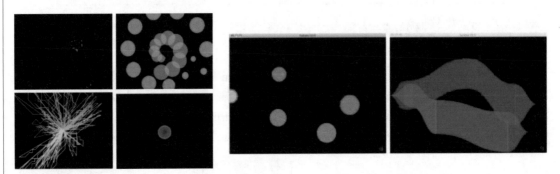

Figure 8.8: Visualizations from left: Spoken Impact Project (Hailpern et al., 2009a) and (B) right: VocSyl (Hailpern et al., 2012).

Another emerging area of voice-based research with autistic individuals is through the use of mobile voice agents like Siri on Apple devices, "OK Google" on Android devices, and Microsoft's Cortana and smart speakers like the Amazon Echo ("Alexa") and Google Home. Most of the work in this space so far has explored how autistic people use these commercially available devices and have been exploratory, such as Allen et al.'s (2018) exploratory study of how autistic people use the Amazon Echo, finding acceptable levels of speech recognition by the devices with improvements upon training. Allen et al. (2018), plus several others (Yu et al., 2018; Parry, 2017), have also explored how Amazon Echo, Amazon Dash, and Amazon Show (which also has a small screen) can be used for speech therapy and visual supports.

8.2.5   FACE, GAZE, AND EYE TRACKING

Technologies that use face or eye tracking have been popular as a means of identifying ways of understanding attention for people who do not speak or may not indicate their attention to researchers in other ways that researchers can easily understand. Likewise, these kinds of techniques have also been used as inputs for teaching dominant culture-socialization skills, such as the reading of facial expressions. As discussed in Chapter 3, Ould Mohamed and colleagues (2006) used gaze tracking and facial orientation to conduct attention analysis in children with autism who were interacting with any type of desktop software. The approach was intended to help children categorize elementary perception, such as whether something is strong, smooth, quick, slow, big, small, etc. This type of approach can be used by other software applications to determine where children attend to in interactive applications, which may improve the design of future applications and interactions.

Another project, Virtual Buddy (Figure 8.9), used eye tracking in very young children with autism aged 24–52 months as a form of intervention in social situations (Trepagnier et al., 2006). The system, which employed a virtual face, monitored whether children engaged in socially appropriate behaviors such as making eye contact, following the face's gaze, and pointing in the same direction. Appropriate behaviors were rewarded with video clips of the child's favorite shows. At the time of publication, the system had undergone pilot testing and was recruiting for a larger study. A similar project used a robot with a lifelike facial display, called FACE, which integrated facial and eye tracking to determine what autistic people were attending to during interactions with the robot. Although the system has only been used with two children, there has been promise demonstrated in this space to conduct therapy and improve social skills (Pioggia et al., 2005).

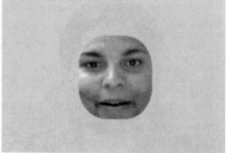

Figure 8.9: **Virtual Buddy system** (Trepagnier et al., 2006).

While the previously described work has enabled tracking of facial, gaze, and eye movements of autistic people themselves, additional work has focused on facial expression analyses of non-autistic peers or conversation partners as a tool for assisting autistic people in normative social interactions with the dominant culture. Researchers who study affective computing (el Kaliouby et al., 2006; Picard, 2009) have looked at using facial expression tracking as a social prosthetic for autistic people, an effort that substantially engaged adolescents and adults in particular. By using a wearable camera, an individual can use it to scan the faces of people with whom they are interacting and use sophisticated algorithms to provide feedback about the emotions the system detects (Madsen et al., 2009b; Bovery et al., 2019).

8.3 CLASSIFICATION APPLIED TO NATURAL USER INTERFACES

In Figure 2.1, we tagged four representative articles as using natural user interfaces as a technology platform. These included the Spoken Impact Project that uses voice (Hailpern et al., 2009a), a smart object called Polipo (Tam et al., 2017), an eye-tracking study (Klin et al., 2002), and the Abaris project (Kientz et al., 2005). Here we describe how each of these fit into the classification scheme defined in Chapter 2, as well as discuss overall trends we observed for technologies making use of the natural user interfaces platform.

Spoken Impact Project (Hailpern et al., 2009a): The Spoken Impact Project visualizes non-speech utterances of non-verbal autistic children. This work uses *audio* as the primary input mechanism and *audio* and *visual* as the output modalities. The domains covered by the project include teaching of *language/communication* skills and also *restrictive* or *repetitive* behavior because it attempts to encourage vocalizations, which can be a type of repetitive behavior.. The goal of this work was primarily to teach skills or encourage new vocalizations, so this was categorized as *intervention/education*. The target end users for this work include the *person with autism* themselves, as well as a *clinician/therapist* who may be working with them. The research studies were primarily conducted in the setting of a *research lab*, though future use could be for a *clinic* or *school*. The publication venue for this project was at the CHI conference, which is a *computing* venue, and autistic individuals were *testers* and *users* in its design. Finally, the studies conducted were of the type *correlational/quasi-experimental*, and the Spoken Impact Project has reached the maturity of a *functional prototype* that is not yet available for public use.

Polipo (Tam et al., 2017): Polipo is a customizable, tangible object that allows therapists to work with autistic children to explore physical movements. The input is entirely *physical* and the output is *audio*, *tactile*, and *visual*. The domains targeted include *cognition* and *sensory/physiological* interactions in a therapy setting for *functional assessment* and as a therapy *intervention*. The target users are the *person with autism* and *therapists*, who are *professionals*, for use in *clinical* settings. Autistic individuals were included as *users* and *testers* in Polipo's design, and the paper was published in *Tangible, Embedded, and Embodied Interaction*, which is a *computing* publishing venue. The exploratory study was *descriptive* in nature with a *functional prototype*.

Eye Tracking (Klin et al., 2002): The article reviewed applies the use of eye-tracking technologies to understanding whether fixations on natural social situations can be predictors of social competence for autistic people. The input in this case is *spatial* due to the following of eye movements, with output being *visual* (in the form of data analyzed). The application was applied to fixation on social situations, and thus the domain is *social/emotional skills*. The short-term goal for this work was a *scientific assessment* of how autistic people respond differential to social stimuli, with the eventual goal for this work being *screening/diagnosis*. The target end users for this work were *researchers* since the data collected was for scientific research, and the setting was also a *research lab*. The work was published in the *Archives of General Psychiatry* journal, which is a *medical* publication venue, and autistic individuals were *users* in this study. The study conducted was *experimental* in nature, and the eye-tracking technology used is *publicly available*.

Abaris (Kientz et al., 2005): The Abaris project used a digital pen and speech recognition, and thus the input mechanisms are *physical* and *audio*, which are connected to videos with *visual* and *audio* output. The therapy supported by Abaris—Discrete Trial Training—primarily has the function of teaching in the domains of *language/communication skills*, *academic skills*, and *life/vocational* skills. The goal of the therapy, and thus Abaris, is for *functional assessment* and *intervention/education* as well as being used for *parent/clinical training* by allowing for reflection on therapy practices. Abaris was developed for target end users of the type *clinician/therapist* and *educator* for both *home* and *school* therapy settings. The work was published originally at the UbiComp conference, which is a *computing* publication venue, and autistic children were included as *testers* in the design of the system. The research study was of the *correlational* or *quasi-experimental* design, and the maturity of Abaris is a *functional prototype* that is not yet available to the public.

Because many of the technologies in this chapter move beyond what may be considered mainstream platforms, most of the applications in this space came from the *computing* publication venue. There were a wide variety of input and output types, which makes sense given the wide variety of modalities that make up natural user interfaces. We also found that a number of the technologies in this space were more *conceptual* or *functional prototypes*, and most were not yet publicly available due to many of the technologies being fairly novel. There did not seem to be many technologies that had *experimental* validation, unless they were comparing the use of a technology by a person with autism to one without. Most studies were *correlational* in nature. We did not see any specific trends for the domain or goals for technologies that used natural user interfaces.

8.4   FUTURE DIRECTIONS

Natural user interfaces used to be a more niche area of computing, requiring expensive hardware, but with the increasing practicality of sensor-based interaction and improved algorithms for speech and gesture recognition, we believe this is an area of much promise, especially as it can account for the wide variety of needs and preferences for interaction that autistic people may express. We expect that much of the future work on technology for autism will be within this space as new sensing and interaction paradigms are developed and established. For example, Microsoft Kinect and similar depth-camera interaction have already made their way into school settings as an interaction platform for autistic students, such as the Lakeside Autism Center in Issaquah, Washington, using Kinect for engaging children in social activities and in physical therapy–like capabilities.[21] The newest versions of depth cameras are including more advanced facial tracking as well[22], and as eye-trackers and facial recognition improve in accuracy, these technologies may become more practical for everyday use. In addition, algorithms supporting eye tracking with commercial built-in webcams are improving such that they can be used without expensive hardware, which will increase the affordability, ubiquity, and access to these devices. Newer tablets are also including more advanced pen-based interactions with styli which are becoming as easy to use as paper, and thus there may be more of a decline in smart-pen and paper technology, though there will likely always be a place for pen and paper in classrooms (Marcu et al., 2013).

Advances in widely available voice-based assistants such as Apple's Siri, Amazon's Alexa, and Microsoft's Cortana have led to more interest in exploring voice-based interactions for a wide variety of people. Anecdotal stories of children with autism engaging in dialog and using them to answer seemingly endless questions have surfaced, and recent research has explored these devices from an accessibility perspective (Pradhan et al., 2018). We expect to see more research into how

---

[21] https://www.seattletimes.com/photo-video/photography/using-kinect-as-a-therapy-tool-for-autism/
[22] http://msdn.microsoft.com/en-us/library/jj130970.aspx

these voice-based assistants will be used for more social development, similar to what we have been seeing in more embodied interactions with social robots, which we describe in Chapter 9.

There is also potential for these types of activities to be adopted beyond just children to adults who may be more comfortable manipulating objects or making gestures than they might be learning to use a more traditional keyboard and mouse. The one downside to these applications is often their complexity in getting set up and teaching people to use and recognize the types of interactions that are allowed by a system. The learnability, as well as the discoverability of new features, of these types of interactions, although "natural," is still somewhat limited and rely on social learning from others, which can be difficult for autistic people. In addition, as computing becomes smaller and more embedded in everyday objects, we will likely see more use of sensors embedded in tangible objects, increasing the reach and scale of tangible computing. Research is always expanding, and we believe will make natural user interfaces more intuitive to learn and interact with, especially for those with autism.

CHAPTER 9

Robotics

In this chapter, we discuss robotics and the use of socially assistive robots (SAR) for autism. While virtual agents and animated toys are sometimes considered robots, this chapter focuses on physical instantiations (both anthropomorphic and humanoid) capable of carrying out behaviorally or socially contingent actions (both autonomously and operated by humans).

9.1 OVERVIEW

Robotics as a field includes a number of subdisciplines ranging from environmental sensing and navigation (Niku, 2001) to object manipulation (Bicchi and Kumar, 2000) and human-robot interaction (Goodrich and Schultz, 2007). The latter forms the bulk of research in the autism space, and is formally referred to as SAR (Feil-Seifer and Mataric, 2005; Tapus et al., 2007). This subfield has its roots in social robotics: robots that interact and communicate with humans or other autonomous physical agents by following social behaviors and rules attached to their roles (Breazeal, 2004; Fong et al., 2003). The cardinal feature of SAR focuses on understanding and developing ways for robots to sense and influence behavior change in humans; it is thus inherently interdisciplinary, drawing heavily from engineering, computer science, psychology, sociology, and related fields.

Most contingent human-robot interactions involve cameras and other sensors embedded in a setting, or in a robot itself, enabling a researcher to control the robot. This control can be instantiated from an adjacent or distant room or within the same room by manipulating hidden controls, using Wizard of Oz (Riek, 2012) techniques.

For many roboticists, the ultimate goal is to have robots sense and contingently interact with humans in a fully autonomous way. However, the underlying research and development required to realize this vision still requires years or even decades of research and is outside the scope of this review. Computational systems are still deeply limited in their ability to sense and understand humans. For example, to date, robots that can "detect emotion" are still largely limited to correlations of phenomena that can sometimes be described as emotions. Thus, although promising approaches are suggested, fully autonomous robots that interact with people, regardless of an autism diagnosis, are not feasible at this time. In this review, we focus on robotics research that may be most applicable in the coming years.

9.2 ROBOTICS AND AUTISM

The majority of research thus far focuses on direct engagement with autistic users, wherein robots enable embodied interactions by simultaneously providing human-like social cues (e.g., waving, smiling) while maintaining object-like simplicity (e.g., limited facial expressions) and behaviorally contingent responses (Dautenhahn and Werry, 2004; Feil-Seifer and Mataric, 2009, 2011; Goan et al., 2006; Piogga et al., 2008, 2005; Robins et al., 2006; Stanton et al., 2008). Robots of this sort create situations for learning and practicing a variety of skills without the risk of social ostracization (Huijnen et al., 2016; Scassellati et al., 2012). Furthermore, observed engagement levels between robots and autistic individuals are high, a promising finding that, as reviewed in greater detail below, is opening up the possibility of using them in a myriad of ways. For the scope of this review, we have applied a simple taxonomy to the applications of SAR in autism: diagnosis, social and emotional interventions, behavioral interventions, development of language and communication skills, and development of motor skills.

9.2.1 DIAGNOSIS

As discussed in Chapter 1, there is no specific biomarker, laboratory test, or neuropsychological assessment procedure to identify autism; it is defined exclusively by behavioral criteria assessed by an experienced and formally trained observer. Scassellati and colleagues (Scassellati, 2005, 2007; Tapus et al., 2007) were some of the first to theorize that SAR can aid in the diagnosis of autism in two primary ways (see Figure 9.1): controlling variables in diagnostic testing and making high-resolution recordings of behavior. Because robots can be programmed to provide consistent actions, diagnosticians could use them to ensure that identical or nearly identical social stimuli are presented to those being evaluated within and across sessions. The goal with such control and replication is to reduce administrative bias in testing (Klin et al., 2000). Robots can also record social behaviors (through embedded or linked sensing modalities) currently rated by trained human observers in situ during test administration or via video recordings at a later time. Common elements to record include head orientation, eye gaze, tone of voice, and object interactions. Taken together, robot-enabled diagnostic capabilities have the potential to support consistency of testing procedures and enable behavior samples to be analyzed in potentially more precise, standardized, and scalable ways. However, it is important to note that robots are not likely to ever replace human diagnosticians and that some autistic people may find such a computational approach dehumanizing and wish for opportunities to provide self-reports of their own experiences when possible. The balance of these competing interests will be essential to the progress of the field long term.

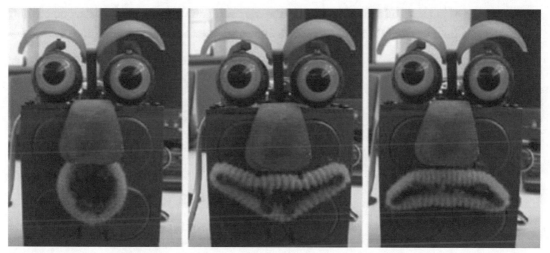

Figure 9.1: Robotic facial expressions of the ERSA robot (Scassellati, 2007).

9.2.2   SOCIAL AND EMOTIONAL INTERVENTIONS

Researchers have explored the use of SAR to enable autistic children to learn, engage in, and enhance social-emotional skills (e.g., Michaud and Théberge-Turmel, 2002), wherein evoking joint attention, eliciting imitation, mediating turn taking, and modeling emotions all provide unique opportunities for social and emotional therapeutic and behavioral interventions.

Frequently, autism is diagnosed due to limitations in joint attention, which refers to the ability to share a common focus of attention either by following a pointing gesture or looking in the same direction another person is looking (Mundy et al., 1990; Tomasello, 1995). In typically developing children, joint attention supports language and social communication (Mundy and Neal, 2001) and has been identified as a key prerequisite for Theory of Mind (Charman et al., 2001). Although a variety of pathways can lead to eventual language development, the centrality of this pathway to typical development has spurred interest in robotic interventions to support joint attention (e.g., Dautenhahn, 2003; Dautenhahn et al., 2009; Da Silva et al., 2009; Feil-Seifer and Mataric, 2009; Kozima et al., 2005, 2007; Piogga et al., 2006; Robins et al., 2009, 2004; Scassellati et al., 2018).

Part of the interest in using robots to elicit joint attention—and one of the major limitations of this approach—is generalizability. In most studies (e.g. Garzotto and Gelsomini, 2016; Bonarini et al., 2016b; Gelsomini et al., 2017), autistic participants engaged in joint attention with robots but not with unfamiliar adults. One might ask then, could robots be used to stimulate joint attention initially and then serve as a scaffold to generalize the skill in human-to-human interactions? Alternately, is this an indication that robots are so fundamentally different from humans that demonstration of joint attention in this context does not ultimately support social development and human-human interaction? Ultimately, this is an empirical question that requires more research.

Figure 9.2: **Left, center:** Keepon robot (Kozima et al., 2007) and **right:** Bandit (Scassellati et al., 2012).

Imitation, another key component of social learning and inter-subjectivity, can be limited in children with autism compared to non-autistic peers (Rogers, 1999; Rogers and Bennetto, 2000). Several studies have shown that autistic children readily imitate robots more than humans (Bird et al., 2007; Boccanfuso and O'Kane, 2011; Duquette et al., 2008; Pierno et al., 2008; Torres et al., 2012). Similar to the aforementioned joint attention findings, there is interest from researchers in seeing if imitation learning with robots could be leveraged and transferred to interactions with humans.

Like joint attention and imitation skills, turn taking is a critical skill needed to mediate social interactions, and can be more challenging for autistic children than for their non-autistic peers. Robots can serve as a powerful point of common interest between two individuals with autism, thus serving as a catalyst and boundary object for social interaction (Costa et al., 2010; Feil-Seifer and Mataric, 2009; Robins et al., 2009, 2005; Wainer et al., 2010; Werry et al., 2001). Given the general interest of researchers, educators, and parents in teaching social skills such as turn taking, there may be room for examining cross-platform and multi-modal approaches to these kinds of skills that include robotics with other technological approaches. To this extent, an example can be found with Teo (Bonarini et al, 2016a; Garzotto and Gelsomini, 2016), a soft-body and mobile robot integrated with multimedia content and virtual worlds presented on projections, use of depth sensors (e.g., Kinect) to enable interaction at a distance, and the virtual elements on display.

Figure 9.3: Triadic interaction between child, robot and display (Bonarini et al, 2016a; Garzotto and Gelsomini, 2016).

It has also been suggested that difficulties perceiving and producing emotions in self and others can give rise to, or exacerbate some of, the diagnostic features of autism. In particular, autistic children can become frustrated working to interpret facial expressions and body language from parents, siblings, and others in their lives. Interacting with people involves substantial sensory stimulation. For autistic individuals who have sensory-processing issues, interacting with another person can create uncertainty and distress (Robins et al., 2009). Child–robot interactions are markedly different in the sense that robots can be programmed to teach basic cause-and-effect elements of social interactions (Kahn et al, 2012). For instance, using embedded sensors, Teo (Bonarini et al, 2016a; Figure 9.4), Kaspar (Wainer et al, 2014; Robins and Dautenhahn, 2014), and Sam (Colombo et al, 2016) can all react contingently to physical interactions such as producing a smile when a user engages in a positive behavior (e.g., caressing the robot) or making crying noises or frowning as a result of negative behaviors (e.g., kicking the robot). Such approaches are often used with typically developing children to teach emotions at a young age using laminated pictures of other children or emojis displaying facial expressions indicative of discrete emotions.

| Child 5 Hits the Robot | Child 6 Creates an Angry Face | ... and Shows it to Child 5 | Child 6 Points Out the Guilty |

Figure 9.4: Communication between children mediated by a robot (Bonarini et al., 2016a)

A small but growing number of researchers have begun to explore whether robots can sense, embody, and teach emotions to individuals with autism using neurological (e.g., EEG), physiological (cardiovascular, electrodermal, etc.), gestural, vocal, eye gaze, and facial expression sensors (Liu et al., 2008a; Mazzei et al., 2010, 2011, 2012). This nascent area of research routinely captures the imagination of the popular press. However, commercial applications of emotion recognition are only as good as a robotic system's ability to provide valid assessments of emotions, which is heavily debated, with the current state of the art typically only engaging in course measures, such as happiness or sadness (e.g., Probo (Pop et al., 2013; Simut et al., 2016)).

Finally, a sizable portion of the SAR literature in autism focuses on entertaining animal-like toys including the baby seal Paro (Marti et al., 2005), the dolphin Sam (Colombo et al, 2016), the teddy bear Huggable (Stiehl et al., 2005), and the robotic cat NeCoRo (Libin and Libin, 2004). The purpose of these robots is often to provide social-emotional support, comfort, or serve therapeutic purposes.

9.2.3   BEHAVIORAL INTERVENTIONS

Reducing maladaptive behaviors and teaching prosocial behaviors is essential to promoting a positive school environment and student achievement of long-term educational and career goals (Ochs et al., 2001, 2004). A common approach to teaching adaptive behaviors involves memorizing and rehearsing social scripts and rules (Cabibihan et al., 2013). These types of interventions lend themselves well to SAR, which are inherently rule-based. In this section, we describe ways in which robotics have been used in various behavioral interventions while respecting the notion of "autistic sociality" (Oches and Solomon, 2010) and other neurodiverse perspectives (Armstrong, 2015).

Many individuals with autism struggle to ask for what they need or to initiate interactions independently. Robots such as Bobus, CPAC, Diskcat, Jumbo, Maestro, Roball (Michaud et al, 2003), CHARLIE (Boccanfuso and O'Kane, 2011), Pekee (Salter et al., 2006), Robota (Billard et al., 2007), Auti (Andreae et al., 2014), Teo (Bonarini et al., 2016a), and Puffy (Gelsomini et al., 2017) have all been tested in therapeutic settings to support autistic users in initiating interactions.

Probolino (Cao et al., 2015) was explicitly designed for home-based therapies. These robots perform an action only after the user has intentionally pressed a button, made a sound, or gestured in a specific way. The resulting action of the robot serves as a reward. Although these scenarios are relatively contrived, there is a possibility to explore how these skills may be transferrable outside the therapy session.

Overcrowded and under-resourced schools and clinics are common as the population of exceptional learners increasingly outnumbers the therapeutic workforce (Aragon, 2016). Remote-controlled robots that can be changed according to a person's abilities and preferences (Ferrari et al., 2009) may be able to address this to a degree, allowing specialists to provide supports remotely without the burden of travel. However, without true robotic autonomy, this kind of engagement cannot easily scale. Thus, researchers have begun to explore the feasibility of robots presenting a sequence of desired emotions without being controlled by the therapist, finding that such approaches are feasible (Gillan, 2019), but the overall notion of "emotion" must be further unpacked.

9.2.4   LANGUAGE AND COMMUNICATION SKILLS

Several studies demonstrate positive effects using robots to promote language development in individuals with autism. For instance, Alemi et al. developed a robot-assisted intervention to teach English words to Persian-speaking seven- to nine-year-olds with autism. English test scores increased and were maintained after two weeks (Alemi et al., 2015). In another study, after a six-week intervention with a robot that involved imitation and games, English-speaking autistic preschoolers with diagnosed speech deficits improved their receptive and expressive communication skills but did not improve their vocabulary (Boccanfuso et al., 2017). More recently, Fuglerud and Solheim (2018) employed an Aldebaran Nao robot (combined with pictures presented on another larger surface) and asked autistic participants to listen to spoken and displayed words and then repeat it. The results of this work indicated that a robot-supported scheme contributed to language learning by combining different means of communication in a systematic way: by association, visualization, listening (to the robot, the session leader, and other children), pronunciation, and repetition. Similarly, Ele, is a plush social elephant that speaks through a live, digitally modified voice of a remote caregiver and augments verbal communication through body movements (such as ears, eyes, and trunk). Eighty percent of children (N=11) demonstrated more eye focus and engagement in conversations with Ele compared with human speakers (Fisicaro et al., 2019; Figure 9.5).

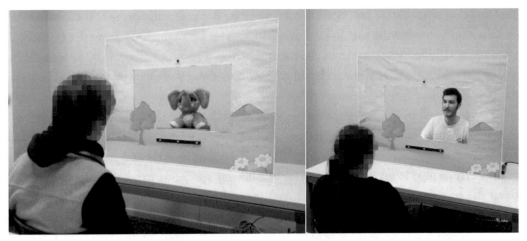

Figure 9.5: **Ele** (Fisicaro et al., 2019)

9.2.5 MOTOR SKILLS

Motor skills encompass a wide range of physical abilities, including fine and gross motor skills, motor fluency, posture, and body awareness. Robots have become increasingly important in a variety of occupational and physical therapy scenarios (e.g., Fasoli 2018; Ríos-Rincón et al., 2016). The advent of force-sensing resistors (FSR)—low-cost and robust sensors which can measure force or pressure—have enabled robotic interventions for fine motor skills and tactile interactions. In a study of eight children with autism and related diagnoses, researchers used Kaspar to simulate gestures involved in social interaction (Costa et al., 2015). Kaspar is minimally expressive and child sized. It can move its torso, arms, head, and invoke limited facial expressions to collectively communicate a limited set of social and emotional responses. The results of their work suggested that Kaspar can be used as an effective tool to elicit new knowledge about body parts, as well as an object of shared attention to promote social interactions with a human partner.

Most robots do not spatially interact with children nor move substantial distances. In some cases, they have slow and clumsy movements that are disturbing or boring, such that children lose attention in movement-related tasks (Shamsuddin et al., 2012). Notable exceptions are IROMEC (Van Den Heuvel et al., 2017), Labo-1 (Dautenhahn and Werry, 2004), Teo (Bonarini et al., 2016a; Garzotto and Gelsomini, 2016), and Puffy (Garzotto et al., 2017a), which all invite exploration of space-related activities. Furthermore, IROMEC (Ferrari et al., 2009) was developed in part to help autistic individuals produce arm and head movements in a simplified way as a building block to later imitation and basic coordinated behaviors.

9.3 CLASSIFICATION SCHEME APPLIED TO ROBOTICS

In Figure 2.1, we classified four different papers as having contributions primarily involved in robotics. This includes Teo (Bonarini et al., 2016a), Behavior-Based Behavior Intervention Architecture (Feil-Seifer and Mataric, 2009), Ele (Fisicaro et al., 2019), and Puffy (Garzotto et al., 2017a). Below we describe how these robotics papers were classified according to the classification scheme defined in Chapter 2.

Teo (Bonarini et al., 2016a): Teo is a plush social robot about 60 cm high. The base consists of a robust, triangular, holonomic, and omnidirectional station that contains motors, batteries, controllers, a colored LED strip, infrared, and sonar sensors. The body, attached to the base, is made of a cloth sack filled with polyester stuffing. A force-sensing resistor (FSR) is sewn in the body cloth to recognize *physical* interactions such as caresses, hugs, and slaps. Thanks to its distance sensors, Teo can *spatially* recognize people moving and change direction accordingly. Teo speaks, moves, and shows colored lights according to its internal behavior (happy, sad, angry, afraid, idle). The goal of the system is to *scientifically assess* interaction between *persons with autism* and *peers* involving them as *users* and *testers* in a clinic. The work has been published at RO-MAN, a computing conference. Empirical support was *experimental* and the system is currently a *functional prototype*.

Behavior-Based Behavior Intervention Architecture (Feil-Seifer and Mataric, 2009): In this research, a robot interacts with and observes an individual with autism using *audio*, *spatial*, and *physical* input modalities and *audio*, *visual*, and *tactile* output modalities. The system observes autistic individuals in terms of their *social/emotional skills* through a combination of video, audio, and physiological sensors on board, embedded in the environment, and/or worn by the individual under observation. The goal of the system is to facilitate *diagnosis/screening*, *intervention/education*, and *scientific assessment* for *persons with autism*, *clinician/therapists*, and *researchers* as target end users. The work involved autistic individuals as *users* and *testers* and was carried out is a *research lab*. It was published in *Experimental Robotics*, a *computing* journal. Empirical support was *experimental* and the system is currently a *functional prototype*.

Ele (Fisicaro et al., 2019): Ele is a plush social robot with an elephant appearance designed as a *conversational* companion to connect *persons with autism*, *peers*, and *professionals* in a *clinic* setting. ELE speaks by streaming the live voice of a remote caregiver, enriching the communication through controlled body movements (*audio* and *physical* output modalities). It is integrated with a tool for automatic gathering and analysis of vocal and facial data that support therapists in monitoring users during their experience with the robotic companion (*audio* input). The work has been published at HRI, a *computing* conference. Empirical support was *experimental* and the system is currently a *functional prototype*.

Puffy (Garzotto et al., 2017a): Puffy is a robotic companion designed in cooperation with a team of therapists and special educators as a learning and play companion for children with autism and their peers. Puffy uses voice, lights, projections embedded in its body, as well as movements in space to display its behaviors (*audio*, *visual* and *spatial* output modalities). It has an egg-shaped, inflatable, soft, and mobile body and can interpret a user's gestures on its body using a force-sensing resistor, and motor movements and facial expressions using a color+depth camera positioned in the head (*spatial* and *physical* input modalities). Children with autism and their peers were assessed in a *research lab* and a *school* as *users* and *testers*. Empirical support was *descriptive* and the system is currently a *functional prototype*. The work was published at CHI and RO-MAN, two *computing* conferences.

The literature reviewed in this area is quite diverse. While robotics is a core feature, several systems also included video and multimedia and sensor-based and wearable sensors. The domain most often targeted was social/emotional, with the goals of scientific assessment and intervention/education. Target end users of these systems tended to focus on researchers with data collection carried out in research labs. Publication venues for the work most often appeared in computing venues, but some also featured in autism-specific journals. Empirical support was mostly descriptive and correlational/quasi-experimental, with few experimental results. Most systems reviewed are currently functional prototypes.

9.4   FUTURE DIRECTIONS

Robots can play several important roles and have potential benefits in therapeutic and educational settings to support skill development, behavioral change, social and emotional development, and support aging in those who identify as autistic. Notwithstanding this promise, there are a variety of current limitations that require more research and development.

As raised in previous reviews (Diehl et al., 2012; Ricks and Colton, 2010; Scassellati et al., 2012), most published robotic studies involving autistic participants still do not systematically evaluate generalizability of robot-mediated outcomes outside the experimental settings. It is thus unclear whether displayed or acquired behaviors and skills transfer to other settings and non-robotic interactions. Such evaluations are critical before investing additional time and money in developing robots to serve as an assistive technology.

All the robotic research included in this review also demonstrated weaknesses recognizing children's speech. Currently there are no autonomous robots that are accurate enough to recognize unpredictable and/or low-volume user pronunciations. Researchers and therapists typically ask autistic learners to pronounce a limited set of words (e.g., up, down, left, right, hello) or use Wizard of Oz to overcome this limitation. Until robots can autonomously understand a wide variety of speech patterns, their use with and by people in their everyday lives will remain limited. This also applies to human users understanding and sustaining interactions with robots. Presently, robotic speech intonation, rhythm, and speed is unsatisfactory, and studies have demonstrated that concentration and engagement are higher in expressive robots compared to robots with flat voices (Westlund et al., 2017). Relatedly, most roboticists follow the guideline of building robots shorter than the people they are going to interact with to reduce distress and anxiety. However, more work can and should be done to ensure that familiarity, harmonic shapes, and appropriate colors are used that fit the culture and context of their use. Only a few robots' designers enable the possibility to change the appearance of the robot to better accommodate user preferences.

While robots appear to engage and elicit various types of clinically significant behaviors in autistic children, the majority of published literature at this time is theoretical and preliminary and focused only on children, far from the level of confirmatory evidence required by health systems, schools, and insurers or to generalize out to a larger group of autistic people. Thus, moving forward, it will be important to attend to the variety of methodological approaches appropriate in the various venues that might publish work on robotics and autism as well as to ensure broad dissemination of these findings in ways that can be accurately interpreted by clinicians, researchers, and technologists.

Another challenge in this area is that robots are expensive and can thus be hard to scale beyond research labs. Commercial grade, low-priced robots that are also safe and robust for a variety of contexts are still not commonplace. At the time of this writing, only a third of the reviewed robots reach a level of maturity to be publicly available at an affordable price (Gelsomini, 2018).

Finally, long-term ethical decisions will have to be agreed upon and norms established regarding the role robots play in diagnosing, socializing, and supporting skill acquisition in individuals with autism. Current expectations have been well described (Coeckelbergh et al., 2016; Peca et al., 2014). However, expectations and what society considers ethical and appropriate tends to evolve with new technologies and contexts. Rather than be simply reactive, researchers in this space should engage proactively with ethicists and philosophers on these important issues.

CHAPTER 10

Discussion and Conclusion

Autism is a complicated diagnosis, label, identity, and concept. Conceived as an umbrella of disabilities by clinicians, autistic experiences as described by those who live them are as vast and varied as any. Autism diagnoses are highly prevalent, engendering interest from autistic people and a vast array of professional caregivers, researchers, educators, politicians, and more. The International Society for Autism Research (INSAR), the largest annual conference on autism, attracts large numbers of interdisciplinary researchers as well as many people who self-identify as autistic. This conference has pushed, alongside designers and computing specialists, a notion that technologies specifically for use by people with autism should be studied. We note as well that technologies designed by people with autism should be included alongside their views, critiques, and insights.

Technologies for autism, much like any technologies, can be used for empowerment, self-actualization, support, and safety, but they can also be used to disempower, oppress, and reinforce stereotypes. At the time of this writing, much concern has been raised globally about the use of technology by bad actors to radicalize terrorists, influence elections, support fascist police states, and more. At the same time, one needs to look no further than the hashtags #ActuallyAutistic and #AbledsAreWeird to see a world of self-expression and empowerment enabled by novel technologies.

The media attention and passion of people with autism and their teachers and families has fueled what some may consider to be a tech bubble in the autism space—myriad applications and devices, all claiming positive results. Overall, the limited evidence as to the efficacy of these technologies is somewhat unsurprising given the limited evidence for other therapeutic approaches that have become popular—e.g., occupational therapy, music therapy, animal therapy, diets and supplements, etc. Similarly, these therapies are often presented in a way that feeds off fear of an autistic life, claiming to "cure" something that for many people there is no reason to cure. Indeed, social media, itself a nascent and at times innovative technology, has become ground zero for discussion of the autistic experience, and sometimes bullying and harassment, of people with autism deemed less fit by their neurotypical counterparts.

This view to inclusion should not belittle those families coping with severe and debilitating forms of autism. However, the overall environment created in the face of such experiences is one in which families may be desperate to try new therapies before their early intervention support runs out. In these cases, waiting for research evidence may not be feasible or desirable, and consideration for neurodiversity may be frightening. Thus, there is a bit of a rush of technology "snake oil" vying for venture capital funding, educational contracts, and parent attention.

Multidisciplinary research in any field struggles with issues of how to conduct research, where to publish, and what constitutes a contribution to the field. In the case of autism and technology, the basic science required to move the field forward includes both the development of novel technologies—some that may be theoretical only or include prototypes not nearly robust enough for regular use—and the construction of theoretical knowledge from empirical research and scientific models. Thus, translating basic science into real-world applications is extraordinarily difficult, because it involves both "tech transfer" and the development of interventions and policies that apply what is known in the scientific space to what is needed in the clinical or educational space. This pipeline requires substantial funding across the way for innovation, technology development, and empirical research, and knowledge about a wide variety of fields. To influence all—or at least multiple—of these fields, researchers must understand and deploy diverse evaluation methods and meet variable standards appropriate to each field.

Additionally, researchers and designers seeking to include the direct perspectives of people with autism must work to create inclusive environments within their research labs. Universities and corporate research labs are often byzantine labyrinths of bureaucratic challenges and hurdles that make it difficult to conduct community-based research or to be autistic and an active student or employee in such a context. Add into that the difficulty of paying people with autism to participate who may need to keep income below a certain point or risk losing benefits, the difficulty of conducting work when transportation is not a given, and other challenges, and it is no wonder that "actually autistic" people are often excluded from the work of design and research. However, it is imperative that we as a community make the effort to change this paradigm.

For the last several years, in the U.S., the Food and Drug Administration (FDA) has been examining whether and how to regulate digital health applications (Shuren et al., 2018). The outcome of this ultimate regulation will likely slow down the development—particularly commercial development—of some technologies for autism. However, the overall effect of this slowdown will also likely be technologies with greater evidence of their efficacy before going to market. One can hope that changing the speed of development may also increase the ability for teams to include more people with autism in their design, development, and testing processes. To meet the high bar of the FDA and other regulatory agencies, the community will have to develop high-quality standards for evaluation. Most of the testing in the autism and technology literature thus far typically observes small numbers of study participants for short periods of time in a single setting. Detailed qualitative analyses, deep understanding of surrounding contexts over long periods of time, and high numbers of subjects in controlled lab settings should each be deployed in the appropriate situations to generate high-quality scholarly output. However, only the latter will ultimately be acceptable for demonstrating a level of statistically significant outcomes required to make claims of therapeutic efficacy.

Similarly, gold standard assessments such as the Autistic Diagnostic Observation, Autism Diagnostic Interview, and Vineland will likely need to be used to confirm diagnostic criteria for autism and substantiate claims that technologies tested are, in fact, impacting the lives of individuals with autism and not just those within the wide range of typical development. Very few studies measure or report on potentially important demographic (age, sex, race, socio-economic status, etc.), functional (e.g., developmental age, IQ, verbal ability, etc.), and confounding factors or associated conditions (e.g., anxiety, medication status, etc.) that may account for observed effects (or lack thereof) and at the very least provide the kind of context needed to interpret results. Furthermore, the heterogeneity in autism makes it very difficult to know whether individual differences within the diagnostic group are driving outcomes. Larger samples of either more diverse or narrowly defined study participants are needed to clarify response types in trials. These studies should also employ a control group or control condition to ascertain whether observed outcomes are attributable to features of study participants or the technology or some other factors. Although we recognize that controlled trials are only one way to determine if interactive technologies are accounting for responses above and beyond other interactions, they are a solid step that is understood by clinicians and provide support for policymakers and administrators looking to implement evidence-based practices. Finally, conducting assessments across time and settings would help determine test-retest reliability, generalizability, and maintenance of observed responses.

While the field must evolve methodologically in the ways described above to support systematic, reliable, regulated interventions, the broad space of research around technologies for autism must also expand to include more high-quality discussion of ethics, humanist research, ethnographic engagement, and the views of people with autism directly. Deeper engagement with participants and more nuanced and careful qualitative observations and analyses can help transfer these results to other people and domains even if they cannot be generalized. Further intentional conversations with and among neurodiverse teams of researchers, community organizers, and advocates will serve not only to improve the quality of the research but will also ensure that the research is sanctioned by and inclusive of the autistic community.

For publishing outlets, technology research typically appears in the proceedings of discipline-specific annual conferences. While highly competitive and peer reviewed, their length and focus necessitate limited description of participant samples and experimental design and more attention to technical and system-level details. In contrast, behavioral science research is typically published in discipline-specific monthly or quarterly journals in longer formats. While also highly competitive and peer reviewed, their length and focus prioritize in-depth descriptions of participant samples, psychometrics of assessments used, experimental design, and interpretation of findings, and less technical detail. As yet another model, medical research is typically published in monthly or quarterly journals in short articles with carefully defined formats.

Each publication outlet has its strengths and weaknesses, and when faced with these articles, researchers from various fields tend to be able to interpret them. However, they are generally not cross-indexed by the various libraries (e.g., PubMed vs. JSTOR vs. ACM/IEEE), and students new to the field may not know to check venues that are less familiar to them. This review cuts across these venues as a starting point for a true multidisciplinary view of the field, and since its initial publication, the classification scheme defined in Chapter 2 has enabled some shared vocabulary across fields and through interdisciplinary teams. However, this review is just one snapshot in time. Technology is rapidly evolving, and the field is maturing.

Additionally, paywalls and the general challenges of academic publishing make it unlikely that the research we describe makes it to the people who can most benefit from it. Thus, we join the chorus of community-based scholars asking for researchers to write blog posts and other public-facing documents explaining their findings and opening up the space for more discussion. Perhaps even more boldly, journals and conferences could support people with autism in drafting their own summaries, responses, and critiques of this work or joining in panels with academics describing their work.

Despite the massive growth, there is substantial room for more work in this space. Greater innovation across various technologies is still possible—and very necessary. Cultural and language barriers as well as socio-economic limits prevent many people from taking advantage of technology-based support. Limited perspectives continue to privilege neurotypical views of the world that ultimately result in designs that are functionally, if not actually, inaccessible or ableist.

When we wrote the first edition of this book, we noted that never before has so much computational power been available in such small and portable devices. This remains as true today as it was then. Devices have only become smaller, more prolific, more powerful, and more connected. Risks and challenges continue to lie against this massive growth. Technology is a great enabler and tends to speed up processes already in place, reifying biases, and heightening inequities. As researchers, designers, advocates, teachers, and students, we must intentionally work against these natural forces. Too often clinicians, parents, teachers, and individuals with autism are seduced by claims of a new piece of software with limited empirical basis or ethical consideration. As these technologies mature, researchers have the moral and ethical responsibility to take up this charge and go beyond the design, development, and preliminary testing of novel innovations. Now is the time to develop measures to determine not only feasibility and efficacy of these interventions, but also to identify the underlying hidden biases, risks, and ethical considerations that must be addressed for a just and inclusive future. Likewise, consumers of new technologies must become savvier and insist that commercial products be tested, evaluated, and subjected to independent scrutiny, that algorithms be fully transparent, and that people with autism be included in design and test teams.

Bibliography

Abdel Karim, A. E. and Mohammed, A. H. (2015). Effectiveness of sensory integration program in motor skills in children with autism. *Egyptian Journal of Medical Human Genetics*, 16(4), 375–380. DOI: 10.1016/j.ejmhg.2014.12.008.

Abowd, G. D. and Mynatt, E. D. (2000). Charting past, present, and future research in ubiquitous computing. *ACM Transactions on Computer-Human Interaction* (TOCHI), 7(1), 29–58. DOI: 10.1145/344949.344988. 119

Abowd, G. D., Atkeson, C. G., Bobick, A., Essa, I. A., MacIntyre, B., Mynatt, E., and Stamer, T. E. (2000). Living laboratories: The future computing environments group at the Georgia Institute of Technology. In G. Szwillus and T. Turner (Eds.), *CHI '00: Extended Abstracts on Human Factors in Computing Systems* (pp. 215–216). New York: ACM Press. DOI: 10.1145/633292.633416. 107

Adjorlu, A., Høeg, E. R., Mangano, L., and Serafin, S. (2017). Daily living skills training in virtual reality to help children with autism spectrum disorder in a real shopping scenario. In *2017 IEEE International Symposium on Mixed and Augmented Reality* (ISMAR-Adjunct), (pp. 294–302). IEEE. DOI: 10.1109/ISMAR-Adjunct.2017.93. 90

Agosta, G., Borghese, L., Brandolese, C., Clasadonte, F., Fornaciari, W., Garzotto, F., Gelsomini, M., Grotto, M., Frà, C., Noferi, D., and Valla, M. (2015). Playful supervised smart spaces (P3S): A framework for designing, implementing and deploying multisensory play experiences for children with special needs. In *2015 Euromicro Conference on Digital System Design* (pp. 158–164). IEEE. DOI: 10.1109/DSD.2015.61.

Ainsworth, M. D. S., Blehar, M. C., Waters, E., and Wall, S. (1978). *Patterns of Attachment: A Psychological Study of the Strange Situation*. Hillsdale, NJ: Erlbaum. 6

Ajodan, E. L., Clark-Whitney, E., Silver, B., Silverman, M. R., Southerland, A., Barnes, E., Dikker, S., Lord, C., Rehg, J. M., Rozga, A., and Jones, R. M. (2019). Increased Eye Contact During Parent-Child Versus Clinician-Child Interactions in Young Children with Autism. https://psyarxiv.com/8c2dm. DOI: 10.31234/osf.io/8c2dm. 111

Albinali, F., Goodwin, M. S., and Intille, S. S. (2009). Recognizing stereotypical motor movements in the laboratory and classroom: A case study with children on the autism spectrum. In *Proceedings of the 11th International Conference on Ubiquitous Computing*. New York: ACM Press, 71–80. DOI: 10.1145/1620545.1620555. 114, 115

Albinali, F., Goodwin, M. S., and Intille, S. S. (2012). Detecting stereotypical motor movements in the classroom using accelerometry and pattern recognition algorithms. *Pervasive and Mobile Computing*, 8, 103–114. DOI: 10.1016/j.pmcj.2011.04.006. 114

Alcantara, P. R. (1994). *Effects of Videotape Instructional Packaging on Purchasing Skills of Children with Autism.* 61(1). Retrieved from http://www.ncbi.nlm. nih.gov/entrez/query.fcgi?db=-pubmed&cmd=Retrieve&dopt=AbstractPlus&list_ uids=4512009269746180833related:4aJ9cobgnT4J. DOI: 10.1177/001440299406100105. 42

Alcorn, A., Pain, H., Rajendran, G., Smith, T., Lemon, O., Porayska-Pomsta, K., Foster, M. E., Avramides, K., Frauenberger, C., and Bernardini, S. (2011). Social communication between virtual characters and children with autism. *Lecture Notes in Computer Science*, 6738, 7–14. DOI: 10.1007/978-3-642-21869- 9_4. DOI: 10.1007/978-3-642-21869-9_4. 89

Alemi, M., Meghdari, A., Basiri, N. M., and Taheri, A. (2015). The effect of applying humanoid robots as teacher assistants to help Iranian autistic pupils learn English as a foreign language. In *International Conference on Social Robotics* (pp. 1–10). Springer, Cham. DOI: 10.1007/978-3-319-25554-5_1. 139

Allen, A. A., Shane, H. C., and Schlosser, R. W. (2018). The Echo™ as a speaker-independent speech recognition device to support children with autism: an exploratory study. *Advances in Neurodevelopmental Disorders*, 2(1), 69–74. DOI: 10.1007/s41252-017-0041-5. 126

Altizer Jr, R., Handman, E., Bayles, G., Jackman, J., Cheng, K., Ritchie, S., Newell, T., and Wright, C. (2018). Choreografish: Co-designing a choreography-based therapeutic virtual reality system with youth who have autism spectrum advantages. In *Proceedings of the 2018 Annual Symposium on Computer-Human Interaction in Play Companion Extended Abstracts* (pp. 381–389). ACM. DOI: 10.1145/3270316.3271541. 92

Alwi, N., Harun, D., and Leonard, J. H. (2015). Clinical application of sensory integration therapy for children with autism. *Egyptian Journal of Medical Human Genetics*, 16(4), 393–394. DOI: 10.1016/j.ejmhg.2015.05.009.

American Psychiatric Association. (2000). *Diagnostic and Statistical Manual on Mental Disorders*, 4th Edition, Text Revision. Washington, DC: Author. 3

American Psychiatric Association. (2013). *Diagnostic and Statistical Manual on Mental Disorders* (DSM-5®). American Psychiatric Association Publishing. DOI: 10.1176/appi. books.9780890425596. 3

American Psychiatric Association. (1993). *Task Force on DSM-IV.* DSM-IV draft criteria. American Psychiatric Pub Incorporated.

Ames, M. G. (2019). *The Charisma Machine: The Life, Death, and Legacy of One Laptop per Child.* MIT Press. 30

Amirabdollahian, F., Robins, B., Dautenhahn, K., and Ji, Z. (2011). Investigating tactile event recognition in child-robot interaction for use in autism therapy. In *Engineering in Medicine and Biology Society, EMBC, 2011 Annual International Conference of the IEEE* (pp. 5347–5351). IEEE. DOI: 10.1109/IEMBS.2011.6091323. 123

Andreae, H., Andreae, P., Low, J., and Brown, D. (2014). A study of auti: a socially assistive robotic toy. *Interaction Design and Children*, 245–248. DOI: 10.1145/2593968.2610463. 138

Annabi, H. and Locke, J. (2019). A theoretical framework for investigating the context for creating employment success in information technology for individuals with autism. *Journal of Management and Organization*, 1–17. DOI: 10.1017/jmo.2018.79. 8

Anzalone, S. M., Tilmont, E., Boucenna, S., Xavier, J., Jouen, A.-L., Bodeau, N., Maharatna, K., Chetouani, M., Cohen, D., and the Michelangelo Study Group. (2014). How children with autism spectrum disorder behave and explore the 4-dimensional (spatial 3D+time) environment during a joint attention induction task with a robot. *Research in Autism Spectrum Disorders*, 8(7), 814–826. DOI: 10.1016/j.rasd.2014.03.002.

Aragon, S. (2016). *Teacher Shortages: What We Know. Teacher Shortage Series. Education Commission of the States.* Education Commission of the States. 139

Armstrong, T. (2015). The myth of the normal brain: Embracing neurodiversity. *AMA Journal of Ethics*, 17(4), 348–352. DOI: 10.1001/journalofethics.2015.17.4.msoc1-1504. 138

Arter, P., Brown, T., Law, M., Barna, J., Fruehan, A., and Fidiam, R. (2018). Virtual Reality: Improving Interviewing Skills in Individuals with Autism Spectrum Disorder. In *Society for Information Technology and Teacher Education International Conference* (pp. 1086–1088). Association for the Advancement of Computing in Education (AACE). 90

Aruanno, B., Garzotto, F., Torelli, E., and Vona, F. (2018). HoloLearn: Wearable Mixed Reality for People with Neurodevelopmental Disorders (NDD). In *Proceedings of the 20th International ACM SIGACCESS Conference on Computers and Accessibility* (pp. 40–51). ACM. DOI: 10.1145/3234695.3236351. 95

Ash, J. S., Berg, M., and Coiera, E. (2004). Some unintended consequences of information technology in health care: the nature of patient care information system-related errors. *Journal of the American Medical Informatics Association*, 11(2), 104–112. DOI: 10.1197/jamia.M1471.

Ayres, A. J. (1972). Types of sensory integrative dysfunction among disabled learners. *American Journal of Occupational Therapy*, 26(1), 13–18. 95

Ayres, A. J. and Tickle, L. S. (1980). Hyper-responsivity to touch and vestibular stimuli as a predictor of positive response to sensory integration procedures by autistic children. *American Journal of Occupational Therapy*, 34(6), 375–381. DOI: 10.5014/ajot.34.6.375. 96

Ayres, K. M. and Langone, J. (2005a). Intervention and instruction with video for students with autism: A review of the literature. *Education and Training in Developmental Disabilities*, 40(2), 183–196. 12, 40

Ayres, K. and Langone, J. (2005b). Evaluation of software for functional skills instruction blending best practice with technology. *Technology in Action*, 1(5), 1–8. 12, 41

Azuma, R. T. (1997). A survey of augmented reality. *Presence*, 6(4), 355–385. DOI: 10.1162/pres.1997.6.4.355. 84

Babu, P. R. K., Oza, P., and Lahiri, U. (2018). Gaze-sensitive virtual reality based social communication platform for individuals with autism. *IEEE Transactions on Affective Computing*, 9(4), 450–462. DOI: 10.1109/TAFFC.2016.2641422. 89

Bai, Z., Blackwell, A. F., and Coulouris, G. (2012). Making Pretense Visible and Graspable: An augmented reality approach to promote pretend play. In *Mixed and Augmented Reality (ISMAR), 2012 IEEE International Symposium on* (pp. 267–268). IEEE. DOI: 10.1109/ISMAR.2012.6402567. 92

Baio, J. (2012). Prevalence of autism spectrum disorders: Autism and developmental disabilities monitoring network, 14 sites, United States, 2008. *Morbidity and Mortality Weekly Report. Surveillance Summaries*, 61(3). Centers for Disease Control and Prevention.

Baio, J., Wiggins, L., Christensen, D. L., Maenner, M. J., Daniels, J., Warren, Z., Kurzius-Spencer, M., Zahorodny, W., Rosenberg, C. R., White, T., Durkin, M. S., Imm, P., Nikolaou, L., Yeargin-Allsopp, M., Lee, L.-C., Harrington, R., Lopez, M., Fitzgerald, R. T., Hewitt, A., Pettygrove, S., Constantino, J. N., Vehorn, A., Shenouda, J., Hall-Lande, J., Braun, K. V. N., and Dowling, N. F. (2018). Prevalence of autism spectrum disorder among children aged 8 years—Autism and developmental disabilities monitoring network, 11 Sites, U.S., 2014. *MMWR Surveillance Summaries*, 67(6), 1. DOI: 10.15585/mmwr.ss6706a1. 3

Baird, G., Charman, T., Cox, A., Cox, A., Baron-Cohen, S., Swettenham, J., and Wheelwright, S. (2001). Current topic: Screening and surveillance for autism and pervasive developmental disorders. *Archives of Disease in Childhood*, 84, 468–475. DOI: 10.1136/adc.84.6.468. 3

Baker, F. and Krout, R. (2009). Songwriting via Skype: An online music therapy intervention to enhance social skills in an adolescent diagnosed with Asperger's Syndrome. *British Journal of Music Therapy*, 23(2), 3–14. DOI: 10.1177/135945750902300202. 39

Baker, R., Bell, S., Baker, E., Holloway, J., Pearce, R., Dowling, Z., Thomas, P., Assey, J., and Ware-ing, L. A. (2001). A randomized controlled trial of the effects of multi-sensory stimulation (MSS) for people with dementia. *British Journal of Clinical Psychology*, 40(1), 81–96. DOI: 10.1348/014466501163508. 95

Bakker, S., Van Den Hoven, E., and Antle, A. N. (2011). MoSo tangibles: evaluating embodied learning. In *Proceedings of the Fifth International Conference on Tangible, Embedded, and Embodied Interaction* (pp. 85–92). ACM. DOI: 10.1145/1935701.1935720. 124, 125

Bao, L. and Intille, S. S. (2004). Activity recognition from user-annotated acceleration data. In A. Ferscha and F. Mattern (Eds.), *Proceedings of PERVASIVE 2004* (Vol. LNCS 3001, pp. 1–17). Berlin: Springer. DOI: 10.1007/978-3-540-24646-6_1. 107

Barakova, E. I., Bajracharya, P., Willemsen, M., Lourens, T., and Huskens, B. (2015). Long-term LEGO therapy with humanoid robot for children with ASD. *Expert Systems*, 32(6), 698–709. DOI: 10.1111/exsy.12098.

Baranek, G. (1999). Autism during infancy: A retrospective video analysis of sensory-motor and social behaviors at 9–12 months of age. *Journal of Autism and Developmental Disorders*, 29:213, 224. DOI: 10.1023/A:1023080005650. 109

Baranek, G. T. (2002). Efficacy of Sensory and Motor Interventions for Children with Autism. *Journal of Autism and Developmental Disorders*, 32(5), 397–422. DOI: 10.1023/A:1020541906063.

Baron-Cohen, S. (1989). The autistic child's theory of mind: The case of specific developmental delay. *Journal of Child Psychology and Psychiatry*, 30, 285–98. DOI: 10.1111/ j.1469-7610.1989.tb00241.x. 5

Baron-Cohen, S., Golan, O., Wheelwright, S. and Hill, J.J. (2004) *Mind Reading: The Interactive Guide to Emotions*. London: Jessica Kingsley Limited. 33, 34

Baron-Cohen, S., Leslie, A. M., and Frith, U. (1985). Does the autistic child have a "theory of mind?" *Cognition*, 21(1), 37–46. DOI: 10.1016/0010-0277(85)90022-8. 33

Baron, M., Groden, J. Groden, G., and Lipsitt, L. (2006.) *Stress and Coping in Autism*. New York: Oxford University Press. DOI: 10.1093/med:psych/9780195182262.001.0001. 33, 113

Barrera, F. J., Violo, R. A., and Graver, E. E. (2007). On the form and function of severe self-injurious behavior. *Behavioral Interventions*, 22, 5–33. DOI: 10.1002/bin.228.

Barrow, W. and Hannah, E. F. (2012). Using computer-assisted interviewing to consult with children with autism spectrum disorders: An exploratory study. *School Psychology International*, 33(4), 450–464. DOI: 10.1177/0143034311429167. 39

Bartoli, L., Corradi, C., Garzotto, F., Gelsomini, M., and Valoriani, M. (2013). Exploring motion-based touchless games for autistic children's learning. In *Proceedings of the 12th international conference on Interaction Design and Children* (pp. 102–111). ACM. DOI: 10.1145/2485760.2485774. 91, 121

Barton, E. E., Reichow, B., Schnitz, A., Smith, I. C., and Sherlock, D. (2015). A systematic review of sensory-based treatments for children with disabilities. *Research in Developmental Disabilities*, 37, 64–80. DOI: 10.1016/j.ridd.2014.11.006.

Bartůňková, S. and Smolová, B. P. (2014). Using Virtual Reality as a Therapeutic Modality for Children with Cerebral Palsy: Aa Review and Synthesis Master thesis Supervisor: Author: (April).

Baskett, C. B. (1996). *The Effect of Live Interactive Video on the Communicative Behavior in Children with Autism.* University of North Carolina at Chapel Hill, Chapel Hill. 31

Bates, E. (1979). T*he Emergence of Symbols: Cognition and Communication in Infancy.* New York: Academic Press. DOI: 10.1016/B978-0-12-081540-1.50013-2. 5

Battocchi, A., Ben-Sasson, A., Esposito, G., Gal, E., Pianesi, F., Tomasini, D., and Zancanaro, M. (2010). Collaborative puzzle game: a tabletop interface for fostering collaborative skills in children with autism spectrum disorders. *Journal of Assistive Technologies*, 4(1), 4–13. DOI: 10.5042/jat.2010.0040. 76

Battocchi, A., Pianesi, F., Tomasini, D., Zancanaro, M., Esposito, G., Venuti, P., and Weiss, P. L. (2009). Collaborative puzzle game: A tabletop interactive game for fostering collaboration in children with autism spectrum disorders (ASD). In *Proceedings of the ACM International Conference on Interactive Tabletops and Surfaces*, 197–204. DOI: 10.1145/1731903.1731940. 76

Bauminger N., Gal E., and Goren-Bar, D. (2007). Enhancing social communication in high-functioning children with autism through a co-located interface. *6th Int. Workshop on Social Intelligence Design*, Trento, 2007. DOI: 10.1109/MMSP.2007.4412808. 77, 79, 88

Bekeli, E. T., Lahiri, U., Swanson, A. R., Crittendon, J. A., Warren, Z. E., and Sarkar, N. (2013). A step towards developing adaptive robot-mediated intervention architecture (ARIA) for children with autism. *IEEE Transactions on Neural Systems and Rehabilitation Engineering*, 21(2), 289–299. DOI: 10.1109/TNSRE.2012.2230188.

Bekele, E., Crittendon, J. A., Swanson, A., Sarkar, N., and Warren, Z. E. (2014). Pilot clinical application of an adaptive robotic system for young children with autism. *Autism*, 18(5), 598–608. DOI: 10.1177/1362361313479454.

Bekele, E., Wade, J., Bian, D., Fan, J., Swanson, A., Warren, Z., and Sarkar, N. (2016). Multimodal adaptive social interaction in virtual environment (MASI-VR) for children with Autism spectrum disorders (ASD). In *2016 IEEE Virtual Reality* (VR) (pp. 121–130). IEEE. DOI: 10.1109/VR.2016.7504695. 88

Bell, G. and Dourish, P. (2006). Yesterday's tomorrows: notes on ubiquitous computing's dominant vision. *Personal and Ubiquitous Computing*, 11(2), 133–143. DOI: 10.1007/ s00779-006-0071-x.

Bellani, M., Fornasari, L., Chittaro, L., and Brambilla, P. (2011). Virtual reality in autism: state of the art. *Epidemiology and Psychiatric Sciences*, 20, 235–238. DOI: 10.1017/ S2045796011000448. 85, 103

Bellini, S. and Akullian, J. (2007). A meta-analysis of video modeling and video self-modeling interventions for children and adolescents with autism spectrum disorders. *Exceptional Children*, 73(3), 264–287. DOI: 10.1177/001440290707300301. 40

Belpaeme, T., Vogt, P., Van den Berghe, R., Bergmann, K., Göksun, T., De Haas, M., kanero, J., Kennedy, J., Küntay, A. C., Oudgenoeg-Paz, O.,Papadopoulos, F., Schodde, T., Verhagen, J., Wallbridge, C. D., Willemsen, B., de Wit, J., Geçkin, V., Hoffmann, L., Kopp, S., Krahmer, E., Mamus, E., Montanier, J-M., Oranç, C., and Pandey, A. K. (2018). Guidelines for designing social robots as second language tutors. *International Journal of Social Robotics*, 1–17. DOI: 10.1007/s12369-018-0467-6.

Ben-Itzchak, E. and Zachor, D. A. (2019). Toddlers to teenagers: Long-term follow-up study of outcomes in autism spectrum disorder. *Autism*, 24(1):41–50. DOI: 10.1177/1362361319840226. 8

Benford, P. and Standen, P. J. (2011). The use of email-facilitated interviewing with higher functioning autistic people participating in a grounded theory study. *International Journal of Social Research Methodology*, 14(5), 353–368. DOI: 10.1080/13645579.2010.534654. 36

Benton, L., Johnson, H.,Ashwin, E.,Brosnan, M., and Grawemeyer, B. (2012). Developing IDEAS: Supporting children with autism within a participatory design team. In *Proceedings of the SIGCHI Conference on Human Factors in Computing Systems* (CHI '12), ACM, New York, 2599–2608. DOI: 10.1145/2207676.2208650. 12

Bereznak, S., Ayres, K. M., Mechling, L. C., and Alexander, J. L. (2012). Video self-prompting and mobile technology to increase daily living and vocational independence for students with autism spectrum disorders. *Journal of Developmental and Physical Disabilities*, 24(3), 269–285. DOI: 10.1007/s10882-012-9270-8. 62

Bernard-Opitz, V., Sriram, N., and Nakhoda-Sapuan, S. (2001). Enhancing social problem solving in children with autism and normal children through computer-assisted instruction. *Journal of Autism and Developmental Disorders*, 31(4), 377–384. DOI: 10.1023/A:1010660502130. 39

Bertrand, J., Mars, A., Boyle, C., Bove, F., Yeargin-Allsopp, M., and Decoufle, P. (2001). Prevalence of autism in the United States population: The Brick Township New Jersey investigation. *Pediatrics*, 108, 1155–1161. DOI: 10.1542/peds.108.5.1155. 3

Bettelheim, B. (1967). *The Empty Fortress*. New York:Free Press. 2

Bhattacharya, A., Gelsomini, M., Pérez-Fuster, P., Abowd, G. D., and Rozga, A. (2015). Designing motion-based activities to engage students with autism in classroom settings. In *Proceedings of the 14th International Conference on Interaction Design and Children* (pp. 69–78). ACM. DOI: 10.1145/2771839.2771847. 122

Bian, D., Wade, J., Swanson, A., Weitlauf, A., Warren, Z., and Sarkar, N. (2019). Design of a Physiology-based Adaptive Virtual Reality Driving Platform for Individuals with ASD. *ACM Transactions on Accessible Computing* (TACCESS), 12(1), 2. DOI: 10.1145/3301498. 90

Bicchi, A. and Kumar, V. (2000). Robotic grasping and contact: A review. In Robotics and Automation, 2000. *Proceedings. ICRA'00. IEEE International Conference on* (Vol. 1, pp. 348–353). IEEE. DOI: 10.1109/ROBOT.2000.844081. 133

Billard, A., Robins, B., Nadel, J., and Dautenhahn, K. (2007). Building robota, a mini-humanoid robot for the rehabilitation of children with autism. *Assistive Technology*, 19(1), 37–49. DOI: 10.1080/10400435.2007.10131864. 138

Bimbrahw, J., Boger, J., and Mihailidis, A. (2012). Investigating the efficacy of a computerized prompting device to assist children with autism spectrum Disorder with activities of daily living. *Assistive Technology*, 24(4), 286–298. DOI: 10.1080/10400435.2012.680661. 62

Bird, G., Leighton, J., Press, C., and Heyes, C. (2007). Intact automatic imitation of human and robot actions in autism spectrum disorders. *Proceedings: Biological Sciences*, 274, 3027–3031. DOI: 10.1098/rspb.2007.1019. 136

Bishop, J. (2003). The Internet for educating individuals with social impairments. *Journal of Computer Assisted Learning*, 19(4), 546–556. DOI: 10.1046/j.0266-4909.2003.00057.x.

Blischak, D. M. and Schlosser, R. W. (2003). Use of technology to support independent spelling by students with autism. *Topics in Language Disorders*, 23(4), 293–304. DOI: 10.1097/00011363-200310000-00005.

Blocher, K. and Picard, R. W. (2002). Affective social quest. In *Socially Intelligent Agents* (pp. 133–140). Springer US. DOI: 10.1007/0-306-47373-9_16. 124

Blumberg, S. J., Bramlett, M. D., Kogan, M. D., Schieve, L. A., Jones, J. R., and Lu, M. C. (2013). Changes in prevalence of parent-reported autism spectrum disorder in school-aged US children: 2007 to 2011–2012. *National Health Statistics Reports*, 65, 1–11. 4

Boccanfuso, L. and O'Kane, J. M. (2011). CHARLIE : An adaptive robot design with hand and face tracking for use in autism therapy. *International Journal of Social Robotics*, 3:337–347. DOI: 10.1007/ s12369-011-0110-2. 136, 138

Boccanfuso, L., Scarborough, S., Abramson, R. K., Hall, A. V., Wright, H. H., and O'Kane, J. M. (2017). A low-cost socially assistive robot and robot-assisted intervention for children with autism spectrum disorder: field trials and lessons learned. *Autonomous Robots*, 41(3), 637–655. DOI: 10.1007/s10514-016-9554-4. 139

Bölte, S., Feineis-Matthews, S., Leber, S., Dierks, T., Hubl, D., and Poustka, F. (2002). The development and evaluation of a computer-based program to test and to teach the recognition of facial affect. *International Journal of Circumpolar Health*, 61. http://creativecommons. org/licenses/by/3.0/. DOI: 10.3402/ijch.v61i0.17503. 33

Bonarini, A., Clasadonte, F., Garzotto, F., and Gelsomini, M. (2015). Blending robots and full-body interaction with large screens for children with intellectual disability. In *Proceedings of the 14th International Conference on Interaction Design and Children - IDC '15* (pp. 351–354). DOI: 10.1145/2771839.2771914.

Bonarini, A., Clasadonte, F., Garzotto, F., Gelsomini, M., and Romero, M. (2016a). Playful interaction with Teo, a mobile robot for children with neurodevelopmental disorders. In *Proceedings of the 7th International Conference on Software Development and Technologies for Enhancing Accessibility and Fighting Info-exclusion - DSAI 2016* (pp. 223–231). DOI: 10.1145/3019943.3019976. 136, 137, 138, 140, 141

Bonarini, A., Garzotto, F., Gelsomini, M., Romero, M., Clasadonte, F., and Yilmaz, A. N. C. (2016b). A huggable, mobile robot for developmental disorder interventions in a multi-modal interaction space. In *25th IEEE International Symposium on Robot and Human Interactive Communication, RO-MAN 2016* (pp. 823–830). DOI: 10.1109/ ROMAN.2016.7745214. 135

Bonarini, A., Garzotto, F., Gelsomini, M., and Valoriani, M. (2014). Integrating human-robot and motion-based touchless interaction for children with intellectual disability. In *Proceedings of the 2014 International Working Conference on Advanced Visual Interfaces - AVI '14* (pp. 341–342). DOI: 10.1145/2598153.2600054.

Bondioli, M., Chessa, S., Narzisi, A., Pelagatti, S., and Piotrowicz, D. (2019, June). Capturing play activities of young children to detect autism red flags. In *International Symposium on Ambient Intelligence* (pp. 71–79). Springer, Cham. DOI: 10.1007/978-3-030-24097-4_9. 115, 116

Bos, D. J., Silver, B. M., Barnes, E. D., Ajodan, E. L., Silverman, M. R., Clark-Whitney, E., Tarpey, T., and Jones, R. M. (2019). Adolescent-specific motivation deficits in autism versus typical development. *Journal of Autism Development Disorders*, 50(1), pp. 364–372. DOI: 10.1007/s10803-019-04258-9. 8

Botts, B. H., Hershfeldt, P. A., and Christensen-Sandfort, R. J. (2008). Snoezelen: Empirical review of product representation. *Focus on Autism and Other Developmental Disabilities*, 23(3), 138–147. DOI: 10.1177/1088357608318949. 96

Boujarwah, F. A. (2012). Facilitating the authoring of multimedia social problem solving skills instructional modules. Georgia Institute of Technology, School of Interactive Computing, Ph.D. thesis. (http://hdl.handle.net/1853/43644). 45, 46

Boujarwah, F., Abowd, G. D., and Arriaga, R. (2012). Socially computed scripts to support social problem solving skills. In *Proceedings of the 2012 ACM Annual Conference on Human Factors in Computing Systems* (CHI '12). ACM, New York, 1987–1996. DOI: 10.1145/2207676.2208343. 42, 45

Boutsika, E. (2014). Kinect in education: A proposal for children with autism. *Procedia Computer Science*, 27, 123–129. DOI: 10.1016/j.procs.2014.02.015. 121

Bovery, M. D. M. J., Dawson, G., Hashemi, J., and Sapiro, G. (2019). A scalable off-the-shelf framework for measuring patterns of attention in young children and its application in autism spectrum disorder. *IEEE Transactions on Affective Computing*. DOI: 10.1109/TAFFC.2018.2890610. 128

Boyd, B. A., Baranek, G. T., Sideris, J., Poe, M. D., Watson, L. R., Patten, E., and Miller, H. (2010). Sensory features and repetitive behaviors in children with autism and developmental delays. *Autism Research*, 3(2), 78–87. DOI: 10.1002/aur.124. 96

Boyd, L. E., Day, K., Stewart, N., Abdo, K., Lamkin, K., and Linstead, E. (2018a). Leveling the Playing Field: Supporting Neurodiversity Via Virtual Realities. *Technology and Innovation*, 20(1–2), 105–116. DOI: 10.21300/20.1-2.2018.105. 87

Boyd, L. E., Gupta, S., Vikmani, S. B., Gutierrez, C. M., Yang, J., Linstead, E., and Hayes, G. R. (2018b). VrSocial: toward immersive therapeutic virtual reality systems for children with autism. In *Proceedings of the 2018 CHI Conference on Human Factors in Computing Systems* (CHI '18) (pp. 21–26). DOI: 10.1145/3173574.3173778.

Boyd, L.A. (2018). Designing and evaluating alternative channels: Visualizing nonverbal communication through AR and VR systems for people with autism. Doctoral Dissertation. University of California, Irvine. 87

Boyd, L. E., Jiang, X., and Hayes, G. R. (2017). ProCom: Designing and evaluating a mobile and wearable system to support proximity awareness for people with autism. In *Proceedings of the 2017 CHI Conference on Human Factors in Computing Systems* (pp. 2865–2877). ACM. DOI:10.1145/3025453.3026014. 93

Boyd, L. E., Rangel, A., Tomimbang, H., Conejo-Toledo, A., Patel, K., Tentori, M., and Hayes, G. R. (2016). SayWAT: Augmenting face-to-face conversations for adults with autism. In *Proceedings of the 2016 CHI Conference on Human Factors in Computing Systems* (pp. 4872–4883). ACM. DOI: 10.1145/2858036.2858215. 93, 112

Boyd, L. E., Ringland, K. E., Haimson, O. L., Fernandez, H., Bistarkey, M., and Hayes, G. R. (2015). Evaluating a collaborative iPad game's impact on social relationships for children with autism spectrum disorder. *ACM Transactions on Accessible Computing* (TACCESS), 7(1), 3. DOI: 10.1145/2751564. 74

Boyd, L.A., McReynolds, C., and Chanin, K. (2013). *The Social Compass Curriculum: A Story-Based Intervention Package for Students with Autism Spectrum Disorders*. Brookes Publishing. 64

Bozgeyikli, E. and Raij, A. (2016). Locomotion in virtual reality for individuals with autism spectrum disorder. *ACM Spatial User Interfaces* (SUI), 33–42. DOI: 10.1145/2983310.2985763.

Bozgeyikli, E., Bozgeyikli, L., Raij, A., Katkoori, S., Alqasemi, R., and Dubey, R. (2016a). *Virtual Reality Interaction Techniques for Individuals with Autism Spectrum Disorder: Design Considerations and Preliminary Results* (Vol. 10272, pp. 127–137). DOI: 10.1007/978-3-319-39516-6_12. 91

Bozgeyikli, L., Bozgeyikli, E., Raij, A., Alqasemi, R., Katkoori, S., and Dubey, R. (2016b). Vocational training with immersive virtual reality for individuals with autism: towards better design practices. In *2016 IEEE 2nd Workshop on Everyday Virtual Reality* (WEVR) (pp. 21–25). IEEE. DOI: 10.1109/WEVR.2016.7859539.

Bozgeyikli, L., Bozgeyikli, E., Clevenger, M., Gong, S., Raij, A., Alqasemi, R., Sundarrao, S., and Dubey, R. (2014). VR4VR: Towards vocational rehabilitation of individuals with disabilities in immersive virtual reality environments. *2014 2nd Workshop on Virtual and Augmented Assistive Technology* (VAAT) (9), 29–34. DOI: 10.1109/VAAT.2014.6799466.

Bozgeyikli, L., Bozgeyikli, E., Raij, A., Alqasemi, R., Katkoori, S., and Dubey, R. (2017a). Vocational rehabilitation of individuals with autism spectrum disorder with virtual reality. *ACM Transactions on Accessible Computing*, 10(2), 1–25. DOI: 10.1145/3046786.

Bozgeyikli, L., Raij, A., Katkoori, S., and Alqasemi, R. (2017b). A survey on virtual reality for training individuals with autism spectrum disorder: Design considerations. *IEEE Transactions on Learning Technologies*, 1382(c), 1–20. DOI: 10.1109/TLT.2017.2739747.

Bozgeyikli, L., Bozgeyikli, E., Katkoori, S., Raij, A., and Alqasemi, R. (2018a). Effects of virtual reality properties on user experience of individuals with autism. *ACM Transactions on Accessible Computing* (TACCESS), 11(4), 22. DOI: 10.1145/3267340. 87

Bozgeyikli, L., Raij, A., Katkoori, S., and Alqasemi, R. (2018b). A survey on virtual reality for individuals with autism spectrum disorder: Design considerations. *IEEE Transactions on Learning Technologies*, 11(2), 133–151. DOI: 10.1109/TLT.2017.2739747. 87

Bradley, R. and Newbutt, N. (2018). Autism and virtual reality head-mounted displays: a state of the art systematic review. *Journal of Enabling Technologies*, 12(3), 101–113. DOI: 10.1108/JET-01-2018-0004. 87

Breazeal, C. (2004). Social interactions in HRI: the robot view. *IEEE Trans. Syst. Man Cybern. Part C*, 34(2):181–86. DOI: 10.1109/TSMCC.2004.826268. 133

Breazeal, C. and Aryananda, L. (2002). Recognition of affective communicative intent in robot-directed speech. *Autonomous Robots*, 12(1), 83–104. DOI: 10.1023/A:1013215010749.

Breazeal, C. and Scassellati, B. (1999). A context-dependent attention system for a social robot. *IJCAI International Joint Conference on Artificial Intelligence*, 2, 1146–1151. DOI: 10.1.1.15.3200.

Bricker, D., Squires, J., Mounts, L., Potter, L., Nickel, R., Twombly, E., and Farrell, J. (1999). *Ages and Stages Questionnaire*. Baltimore, MD: Paul H. Brookes. 64

Broaders, S. C., Cook, S. W., Mitchell, Z., and Goldin-Meadow, S. (2007). Making children gesture brings out implicit knowledge and leads to learning. *Journal of Experimental Psychology: General*, 136(4), 539. DOI: 10.1037/0096-3445.136.4.539. 121

Bronfenbrenner, U. (1979). *The Ecology of Human Development*. Cambridge, MA: Harvard University Press. 9

Brown, D. J., Kerr, S. J., and Eynon, A. (1997). New advances in VEs for people with special needs. *Ability*, 19. 87

Brown, S. A. and Gemeinboeck, P. (2018). Sensory conversation: An interactive environment to augment social communication in autistic children. In *Assistive Augmentation* (pp. 131–150). Springer, Singapore. DOI: 10.1007/978-981-10-6404-3_8. 100

Brown, S. A., Silvera-Tawil, D., Gemeinboeck, P., and McGhee, J. (2016). The case for conversation: A design research framework for participatory feedback from autistic children. In *Proceedings of the 28th Australian Conference on Computer-Human Interaction* (pp. 605–613). ACM. DOI: 10.1145/3010915.3010934. 100

Brown, S. and Koh, J. T. K. V. (2014). Responsive multisensory environments as a tool to facilitate social engagement in children with an autism spectrum disorder. *SIGGRAPH Asia 2014 Designing Tools For Crafting Interactive Artifacts on - SIGGRAPH ASIA '14*, 1–4. DOI: 10.1145/2668947.2668949.

Bruner, J. S. (1975). From communication to language: A psychological perspective. *Cognition*, 3, 255–287. DOI: 10.1016/0010-0277(74)90012-2. 5

Brunswik, E. (1947). *Systematic and Representative Design of Psychological Experiments, with Results in Physical and Social Perception*. Berkeley: University of California Press. 108

Bryson, S. E., Clark, B. S., and Smith, I. M. (1988). First report of a Canadian epidemiological study of autistic syndromes. *Journal of Child Psychology and Psychiatry*, 29, 433–445. DOI: 10.1111/j.1469-7610.1988.tb00735.x. 3

Bundy, A. C., Lane, S. J., and Murray, E. A. (2002). *Sensory Integration: Theory and Practice*. FA Davis. 95

Burdea, G. and Coiffet, P. (2003). *Virtual Reality Technology*. Hoboken, NJ:Wiley. DOI: 10.1162/105474603322955950. 84

Burke, M., Kraut, R., and Williams, D. (2010). Social use of computer-mediated communication by adults on the autism spectrum. In *Proceedings of the 2010 ACM Conference on Computer Supported Cooperative Work* (pp. 425–434). ACM. DOI: 10.1145/1718918.1718991. 12, 36, 38, 53

Burke, R. V., Allen, K. D., Howard, M. R., Downey, D., Matz, M. G., and Bowen, S. L. (2013). Tablet-based video modeling and prompting in the workplace for individuals with autism. *Journal of Vocational Rehabilitation*, 38(1), 1–14. DOI: 10.3233/JVR-120616. 57

Burke, S. L., Bresnahan, T., Li, T., Epnere, K., Rizzo, A., Partin, M., Ahlness, R. M., and Trimmer, M. (2018). Using virtual interactive training agents (ViTA) with adults with autism and other developmental disabilities. *Journal of Autism and Developmental Disorders*, 48(3), 905–912. DOI: 10.1007/s10803-017-3374-z. 88

Buzio, A., Chiesa, M., and Toppan, R. (2017). Virtual reality for special educational needs. *Proceedings of the 2017 ACM Workshop on Intelligent Interfaces for Ubiquitous and Smart Learning - SmartLearn '17*, 7–10. DOI: 10.1145/3038535.3038541.

Cabibihan, J. J., Javed, H., Aldosari, M., Frazier, T., and Elbashir, H. (2017). Sensing technologies for autism spectrum disorder screening and intervention. *Sensors*, 17(1), 46. DOI: 10.3390/s17010046. 117

Cabibihan, J. J., Javed, H., Ang, M., and Aljunied, S. M. (2013). Why robots? A survey on the roles and benefits of social robots in the therapy of children with autism. *International Journal of Social Robotics*, 5(4), 593–618. DOI: 10.1007/s12369-013-0202-2. 138

Cai, Y., Chia, N. K., Thalmann, D., Kee, N. K., Zheng, J., and Thalmann, N. M. (2013). Design and development of a virtual dolphinarium for children with autism. *IEEE Transactions on Neural Systems and Rehabilitation Engineering*, 21(2), 208–217. DOI: 10.1109/TNSRE.2013.2240700. 91

Campbell, D. T. and Stanley, J. C. (1966). *Experimental and Quasi-Experimental Designs for Research*. Chicago: Rand McNally. 108

Cannella-Malone, H. I., DeBar, R. M., and Sigafoos, J. (2009). An examination of preference for augmentative and alternative communication devices with two boys with significant intellectual disabilities. *Augmentative and Alternative Communication*, 25(4), 262-273. DOI: 10.3109/07434610903384511. 57

Cao, H. L., Pop, C., Simut, R., Furnemónt, R., De Beir, A., Van de Perre, G., Esteban, P. G., Lefeber, D., and Vanderborght, B. (2015). Probolino: A portable low-cost social device for home-based autism therapy. In *International Conference on Social Robotics*, (pp. 93–102). Springer, Cham. DOI: 10.1007/978-3-319-25554-5_10. 139

Capps, L., Sigman, M., and Mundy, P. (1994). Attachment security in children with autism. *Dvelopment and Psychopathology*, 6, 249–262. DOI: 10.1017/S0954579400004569. 6

Carlile, K. A., Reeve, S. A., Reeve, K. F., and DeBar, R. M. (2013). Using activity schedules on the iPod touch to teach leisure skills to children with autism. *Education and Treatment of Children*, 36(2), 33–57. DOI: 10.1353/etc.2013.0015. 62

Carter, M. and Stephenson, J. (2012). The use of multi-sensory environments in schools servicing children with severe disabilities. *Journal of Developmental and Physical Disabilities*, 24(1), 95–109. DOI: 10.1007/s10882-011-9257-x.

Casas, X., Herrera, G., Coma, I., and Fernández, M. (2012). A kinect-based augmented reality system for individuals with autism spectrum disorders. In *GRAPP/IVAPP* (pp. 440–446). DOI: 10.5220/0003844204400446. 93

Case-Smith, J., Weaver, L. L., and Fristad, M. A. (2015). A systematic review of sensory processing interventions for children with autism spectrum disorders. *Autism*, 19(2), 133–148. DOI: 10.1177/1362361313517762.

Chakrabarti, S. and Fombonne, E. (2001). Pervasive developmental disorders in preschool children. *Journal of the American Medical Association*, 285, 3093–3099. DOI: 10.1001/jama.285.24.3093. 3

Chan, S. W. C., Chien, W. T., and To, M. Y. F. (2007). An evaluation of the clinical effectiveness of a multisensory therapy on individuals with learning disability. *Hong Kong Medical Journal*, 13(1), 28–31. 97

Chan, S., Fung, M. Y., Tong, C. W., and Thompson, D. (2005). The clinical effectiveness of a multisensory therapy on clients with developmental disability. *Research in Developmental Disabilities*, 26(2), 131–142. DOI: 10.1016/j.ridd.2004.02.002.

Chang, K.H., Lei, H., and Canny, J. (2011). Improved Classification of Speaking Styles for Mental Health Monitoring using Phoneme Dynamics. In: *Proceedings of the 12th Annual Conference of the International Speech Communication Association* (Interspeech). Florence, Italy.

Chang, Y. C., Chen, C. H., Huang, P. C., and Lin, L. Y. (2019). Understanding the characteristics of friendship quality, activity participation, and emotional well-being in Taiwanese adolescents with autism spectrum disorder. *Scandinavian Journal of Occupational Therapy*, 26(6), 452–462. DOI: 10.1080/11038128.2018.1449887. 8

Chapple, D. (2011). The evolution of augmentative communication and the importance of alternate access. *Perspectives on Augmentative and Alternative Communication*, 20(1), 34–37. DOI: 10.1044/aac20.1.34.

Charitos, D., Karadanos, E. G., Sereti, E., Triantafillou, Koukouvinou, S., and Martakos, D. (2000). Employing virtual reality for aiding the organisation of autistic children behaviour in everyday tasks, *Proceedings of the 3rd International Conference on Disability, Virtual Reality and Associated Technologies* (ICDVRAT), Sardinia, September 23-25th, 147–152. 87

Charlop-Christy, M. H., Le, L., and Freeman, K. A. (2000). A comparison of video modeling with in vivo modeling for teaching children with autism. *Journal of Autism and Developmental Disorders*, 30, 537–552. DOI: 10.1023/A:1005635326276. 42

Charman, T., Baron-Cohen, S., Swettenham, J., Baird, G., Cox, A., and Drew, A. (2001). Testing joint attention, imitation, and play as infancy precursors to language and theory of mind. *Cognitive Development*, 15, 481–498. DOI: 10.1016/S0885-2014(01)00037-5. 135

Chaspari, T., Provost, E. M., Katsamanis, A., and Narayanan, S. (2012). An acoustic analysis of shared enjoyment in ECA interactions of children with autism. In *Acoustics, Speech and*

Signal Processing (ICASSP), 2012 IEEE International Conference on (pp. 4485– 4488). IEEE. DOI: 10.1109/ICASSP.2012.6288916. 112

Chen, C. H., Lee, I. J., and Lin, L. Y. (2016). Augmented reality-based video-modeling storybook of nonverbal facial cues for children with autism spectrum disorder to improve their perceptions and judgments of facial expressions and emotions. *Computers in Human Behavior*, 55, 477-485. DOI: 10.1016/j.chb.2015.09.033. 92

Chen, S. H. and Bernard-Opitz, V. (1993). Comparison of personal and computer-assisted instruction for children with autism. *Mental Retardation*, 31, 368–376. 39

Chen, W. (2012). Multitouch tabletop technology for people with autism spectrum disorder: A review of the literature. *Procedia Computer Science*, 14, 198–207. DOI: 10.1016/j.procs.2012.10.023. 57, 69

Cheng, L., Kimberly, G., and Orlich, F. (2002). Kidtalk: Online therapy for asperger's syndrome. Microsoft Research. 32

Cheng, Y. and Ye, J. (2010). Exploring the social competence of students with autism spectrum conditions in a collaborative virtual learning environment—The pilot study. *Computers and Education*, 54, 1068–1077. DOI: 10.1016/j.compedu.2009.10.011. 89

Cheng, Y., Huang, C. L., and Yang, C. S. (2015). Using a 3D immersive virtual environment system to enhance social understanding and social skills for children with autism spectrum disorders. *Focus on Autism and Other Developmental Disabilities*, 30(4), 222–236. DOI: 10.1177/1088357615583473. 88

Chia, F. Y. and Saakes, D. (2014). Interactive training chopsticks to improve fine motor skills. In *Proceedings of the 11th Conference on Advances in Computer Entertainment Technology*, (p. 57). ACM. DOI: 10.1145/2663806.2663816. 124, 125

Chin, I., S., Potrzeba, E., Goodwin, M. S., and Naigles, L. (2013a). Verb Use in a Child Previously Diagnosd with ASD: Dense Recordings Reveal Typical and Atypical Development. Poster presented at Society for Research in Child Development 2013 Biennial Meeting, Seattle, WA, April 18th-20th.

Chin, I., Vosoughi, S., Goodwin, M. S., Roy, D. K., and Naigles, L. (2013b). Dense Data Collection Through the Speechome Recorder Better Reveals Developmental Trajectories. Poster presented at International Meeting for Autism Research, San Sebastian, Spain, May 2–4.

Chin, I., Goodwin, M. S., Vosoughi, S., Roy, D., and Naigles, L. R. (2018). Dense home-based recordings reveal typical and atypical development of tense/aspect in a child with delayed language development. *Journal of Child Language*, 45(1), 1–34. DOI: 10.1017/ S0305000916000696. 110

Cho, S. and Ahn, D. H. (2016). *Socially Assistive Robotics in Autism Spectrum Disorder*, 17–26. https://doi.org/10.7599/hmr.2016.36.1.17.

Christinaki, E., Vidakis, N., and Triantafyllidis, G. A. (2014). A novel educational game for teaching emotion identification skills to preschoolers with autism diagnosis. *Computer Science and Information Systems*, 11(2), 723–743. DOI: 10.2298/CSIS140215039C. 59

Cibrian, F. L., Peña, O., Ortega, D., and Tentori, M. (2017). BendableSound: An elastic multisensory surface using touch-based interactions to assist children with severe autism during music therapy. *International Journal of Human-Computer Studies*, 107, 22–37. DOI: 10.1016/j.ijhcs.2017.05.003. 99

Cihak, D., Fahrenkrog, C., Ayres, K. M., and Smith, C. (2009). The use of video modeling via a video iPod and a system of least prompts to improve transitional behaviors for students with autism spectrum disorders in the general education classroom. *Journal of Positive Behavior Interventions*, 12(2), 103–115. DOI: 10.1177/1098300709332346. 42

Ciptadi, A., Goodwin, M. S., and Rehg, J. M. (2014). Movement pattern histogram for action recognition and retrieval. In *European Conference on Computer Vision* (pp. 695–710). Springer, Cham. DOI: 10.1007/978-3-319-10605-2_45. 111

Clark, M. L., Austin, D. W., and Craike, M. J. (2015). Professional and parental attitudes toward iPad application use in autism spectrum disorder. *Focus on Autism and Other Developmental Disabilities*, 30(3), 174–181. DOI: 10.1177/1088357614537353. 57

Cobb, S., Beardon, L., Eastgate, R., Glover, T., Kerr, S., Neale, H., et al. (2002). Applied virtual environments to support learning of social interaction skills in users with Asperger's Syndrome. *Digital Creativity*, 13(1), 11. DOI: 10.1076/digc.13.1.11.3208. 88

Coeckelbergh, M., Pop, C., Simut, R., Peca, A., Pintea, S., David, D., and Vanderborght, B. (2016). A survey of expectations about the role of robots in robot-assisted therapy for children with ASD: Ethical acceptability, trust, sociability, appearance, and attachment. *Science and Engineering Ethics*, 22(1), 47–65. DOI: 10.1007/s11948-015-9649-x. 143

Cohen, I. L., Yoo, H. Y., Goodwin, M. S., and Moskowitz (2011). Assessing challenging behaviors in autism spectrum disorders: Prevalence, rating scales, and autonomic indicators. In J. Matson and P. Sturmey (Eds.) *International Handbook of Autism and Pervasive Developmental Disorders*. (pp. 247–270). Springer. DOI: 10.1007/978-1-4419-8065-6_15. 114

Colby, K. M. (1973). The rationale for computer-based treatment of language difficulties in non-speaking autistic children. *Journal of Autism and Childhood Schizophrenia*, 3(3), 254–260. DOI: 10.1007/BF01538283. 11

Colby, K. M. and Smith, D. C. (1971). Computers in the treatment of non-speaking autistic children. *Current Psychiatric Therapies*, 11, 1. 11

Coleman-Martin, M. B., Heller, K. W., Cihak, D. F., and Irvine, K. L. (2005). Using computer-assisted instruction and the nonverbal reading approach to teach word identification. *Focus on Autism and Other Developmental Disabilities*, 20(2), 80–90. DOI: 10.1177/10883576050200020401. 39

Collier, L. and Truman, J. (2008). Exploring the multi-sensory environment as a leisure resource for people with complex neurological disabilities. *NeuroRehabilitation*, 23(4), 361–367. DOI: 10.3233/NRE-2008-23410. 97

Colombo, S., Garzotto, F., Gelsomini, M., Melli, M., and Clasadonte, F. (2016). Dolphin Sam: A smart pet for children with intellectual disability. *Proceedings of the International Working Conference on Advanced Visual Interfaces - AVI '16*, 352–353. DOI: 10.1145/2909132.2926090. 137, 138

Constantin, A., Johnson, H., Smith, E., Lengyel, D., and Brosnan, M. (2017). Designing computer-based rewards with and for children with Autism Spectrum Disorder and/or intellectual disability. *Computers in Human Behavior*, 75, 404–414. DOI: 10.1016/j.chb.2017.05.030. 40

Constantin, A., Korte, J., Fails, J. A., Good, J., Alexandru, C. A., Dragomir, M., Pain, H., Hourcade, J. P., Eriksson, E., Waller, A., and Garzotto, F. (2019). Pushing the boundaries of participatory design with children with special needs. In *Proceedings of the 18th ACM International Conference on Interaction Design and Children* (pp. 697–705). DOI: 10.1145/3311927.3325165. 12

Cosentino, G., Gianotti, M., Gelsomini, M., Garzotto, F., and Arquilla, V. (2019). Perform the Magic! Usability testing for Magika, a Multisensory Environment fostering children's well being (No. 1376). EasyChair.

Cosentino, G., Leonardi, G., Gelsomini, M., Spitale, M., Gianotti, M., Garzotto, F., and Arquilla, V. (2019). GENIEL: an auto-generative intelligent interface to empower learning in a multi-sensory environment. In *IUI Companion* (pp. 27–28). DOI: 10.1145/3308557.3308685.

Costa, A. P., Steffgen, G., and Samson, A. C. (2017). Expressive incoherence and alexithymia in autism spectrum disorder. *Journal of Autism and Developmental Disorders*, 47, 1659–1672. DOI: 10.1007/s10803-017-3073-9. 113

Costa, S., Lehmann, H., Dautenhahn, K., Robins, B., and Soares, F. (2015). Using a humanoid robot to elicit body awareness and appropriate physical interaction in children with autism. *International Journal of Social Robotics*, 7(2), 265–278. DOI: 10.1007/s12369-014-0250-2. 140

Costa, S., Santos, C., Soares, F., Ferreira, M., and Moreira, F. (2010). Promoting interaction amongst autistic adolescents using robots. *32nd Annual International Xonference of the IEEE/EMBS* (pp. 3856–3859). DOI: 10.1109/IEMBS.2010.5627905. 136

Costa, S., Soares, F., Santos, C., Ferreira, M., Moreira, F., Pereira, A., and Cunha, F. (2011). An approach to promote social and communication behaviors in children with autism spectrum disorders: Robot based intervention. *Proceedings of the 20th IEEE International Symposium on Robot and Human Interaction Communciation*, 101–106. DOI: 10.1109/ROMAN.2011.6005244.

Cramer, M. and Hayes, G. (2010). Acceptable use of technology in schools: risks, policies, and promises. *Pervasive Computing, IEEE*, 9(3), 37–44. DOI: 10.1109/MPRV.2010.42. 30, 72

Cramer, M., Hirano, S. H., Tentori, M., Yeganyan, M. T., and Hayes, G. R. (2011). Classroom-based assistive technology: collective use of interactive visual schedules by students with autism. In *Proceedings of the 2011 Annual Conference on Human Factors in Computing Systems - CHI '11* (p. 1). New York: ACM Press. DOI: 10.1145/1978942.1978944. 71

Crosier, J. K., Cobb, S., and Wilson, J. R. (2002). Key lessons for the design and integration of virtual environments in secondary science. *Computers and Education*, 38(1-3), 77–94. DOI: 10.1016/S0360-1315(01)00075-6. 87, 88

Crowell, C., Sayis, B., Bravo, A., and Paramithiotti, A. (2018). GenPlay: Generative playscape. In *Extended Abstracts of the 2018 CHI Conference on Human Factors in Computing Systems* (p. SDC01). ACM. DOI: 10.1145/3170427.3180653. 99

Cumming, T. M. (2010). Using technology to create motivating social skills lessons. *Intervention in School and Clinic*, 45(4), 242–250. DOI: 10.1177/1053451209353445. 45

Cunningham, D., McMahon, H., ONeill, B., Quinn, A., Loughrey, D., Farren, S., and Creighton, N. (1992). Bubble dialog-A new tool for instruction and assessment. *Etr&D-Educational Technology Research And Development*, 40(2), 59–67. DOI: 10.1007/ BF02297050. 37

Da Silva, P. R. S., Tadano, K., Saito, A., Lambacher, S. G., and Higashi, M. (2009). Therapeutic-assisted robot for children with autism. *IEEE/RSJ International Conference on Intelligent Robots and Systems*, (pp. 3561–3567). New York:ACM Press. DOI: 10.1109/IROS.2009.5354653. 135

Dalton, N. S. (2013). Neurodiversity and HCI. In *CHI'13 Extended Abstracts on Human Factors in Computing Systems* (pp. 2295–2304). ACM. DOI: 10.1145/2468356.2468752. 10

Dattolo, A. and Luccio, F. L. (2016). A review of websites and mobile applications for people with autism spectrum disorders: Towards shared guidelines. In *International Conference on Smart Objects and Technologies for Social Good* (pp. 264–273). Springer, Cham. DOI: 10.1007/978-3-319-61949-1_28. 65

Dauphin, M., Kinney, E. M., Stromer, R., and Koegel, R. L. (2004). Using video-enhanced activity schedules and matrix training to teach sociodramatic play to a child with autism. *Journal of Positive Behavior Interventions*, 6(4), 238–250. DOI: 10.1177/10983007040060040501. 44

Dautenhahn, K. (2003). Roles and functions of robots in human society: Implications from research in autism therapy. *Robotica*, 21, 443–452. DOI: 10.1017/S0263574703004922. 135

Dautenhahn, K. (2007). Socially intelligent robots: dimensions of human-robot interaction. *Philosophical Transactions of the Royal Society B: Biological Sciences*, 362(1480), 679–704. DOI: 10.1098/rstb.2006.2004.

Dautenhahn, K. and Werry, I. (2004). Toward interactive robots in autism therapy: Background, motivation, and challenges. *Pragmatics and Cognition*, 12, 1–35. DOI: 10.1075/pc.12.1.03dau. 134, 140

Dautenhahn, K., Nehaniv, C., Walters, M. L., Robins, B., Kose-Bagci, H., Mirza, N. A., and Blow, M. (2009). KASPAR—a minimally expressive humanoid robot for human-robot interaction research. *Applied Bionics and Biomechanics*, 6(3–4):369–97. DOI: 10.1080/11762320903123567. 135

Dawson, G. and Osterling, J. (1997). Early intervention in autism. In M.J. Guralnick (Ed.), *The Effectiveness of Early Intervention* (pp. 307–326). Baltimore, MD: Paul H. Brooks. 9

Dawson, G., Meltzoff, A.N., Osterling, J., Rinaldi, J., and Brown, E. (1998). Children with autism fail to orient to naturally occurring social stimuli. *Journal of Autism and Developmental Disorders*, 28, 479–485. DOI: 10.1023/A:1026043926488. 5

De Leo, G. and Leroy, G. (2008). Smartphones to facilitate communication and improve social skills of children with severe autism spectrum disorder: special education teachers as proxies. In *Proceedings of the 7th international conference on Interaction Design and Children* (45–48). ACM. DOI: 10.1145/1463689.1463715. 57

de Urturi, Z. S., Zorrilla, A. M., and Zapirain, B. G. (2012). A serious game for android devices to help educate individuals with autism on basic first aid. In *Distributed Computing and Artificial Intelligence* (pp. 609–616). Springer, Berlin, Heidelberg. DOI: 10.1007/978-3-642-28765-7_74. 62

Delano, M. E. (2007). Video modeling interventions for individuals with autism. *Remedial and Special Education*, 28(1), 33–42. DOI: 10.1177/07419325070280010401. 40

De Leo, G., Gonzales, C. H., Battagiri, P., and Leroy, G. (2011). A smart-phone application and a companion website for the improvement of the communication skills of children with autism: clinical rationale, technical development and preliminary results. *Journal of Medical Systems*, 35(4), 703-711. DOI: 10.1007/s10916-009-9407-1. 58

DeMyer, M. K. (1979). *Parents and Children in Autism*. Washington, DC: Victor Winston and Sons. 8

Dickie, C., Vertegaal, R., Shell, J. S., Sohn, C., Cheng, D., and Aoudeh, O. (2004). Eye contact sensing glasses for attention-sensitive wearable video blogging. In *Proceedings of the Conference on Human Factors in Computing Systems: Extended Abstracts* (pp. 769–770). New York: ACM Press. DOI: 10.1145/985921.985927.

Didehbani, N., Allen, T., Kandalaft, M., Krawczyk, D., and Chapman, S. (2016). Virtual reality social cognition training for children with high functioning autism. *Computers in Human Behavior*, 62, 703–711. DOI: 10.1016/j.chb.2016.04.033. 88

Diehl, J. J., Schmitt, L. M., Villano, M., and Crowell, C. R. (2012). The clinical use of robots for individuals with autism spectrum disorders: a critical review. *Research in Autism Spectrum Disorders*, 6(1):249–262. DOI: 10.1016/j.rasd.2011.05.006. 142

Dietz, P. and Leigh, D. (2001). DiamondTouch: A multi-user touch technology. In *Proceedings of UIST 2001*, 219–226. DOI: 10.1145/502348.502389. 69

Dillon, G. and Underwood, J. (2012). Computer mediated imaginative storytelling in children with autism. *International Journal of Human-Computer Studies*, 70(2), 169–178. DOI: 10.1016/j.ijhcs.2011.10.002. 37

Drain, J., Riojas, M., Lysecky, S., and Rozenblit, J. (2011). First experiences with eBlocks as an assistive technology for individuals with autistic spectrum condition. In *Frontiers in Education Conference* (FIE), 2011 (pp. T3D–1). IEEE. DOI: 10.1109/FIE.2011.6143050. 124

Druin, A. (2002). The role of children in the design of new technology. *Behaviour and Information Technology*, 21(1), 1–25. DOI: 10.1080/01449290110108659. 24

Duffield, T. C., Parsons, T. D., Landry, A., Karam, S., Otero, T., Mastel, S., and Hall, T. A. (2018). Virtual environments as an assessment modality with pediatric ASD populations: a brief report. *Child Neuropsychology*, 24(8), 1129–1136. DOI: 10.1080/09297049.2017.1375473.

Duggal, C., Dua, B., Chokhani, R., and Sengupta, K. (2019). What works and how: Adult learner perspectives on an autism intervention training program in India. *Autism*, DOI: 10.1177/1362361319856955. 9

Dunlap, G. (1999). Consensus, engagement, and family involvement for young children with autism. *Journal of the Association for Persons with Severe Handicaps*, 24, 222–225. DOI: 10.2511/rpsd.24.3.222. 9

Duquette, A., Michaud, F., and Mercier, H. (2008). Exploring the use of a mobile robot as an imitation agent with children with low functioning autism. *Autonomous Robots – Special Issue on Socially Assistive Robotics*, 24, 147–157. DOI: 10.1007/s10514-007-9056-5. 136

Dyches, T. T., Wilder, L. K., and Obiakor, F. E. (2001). Autism: Multicultural perspectives. In T. Wahlberg, F. Obiakor, S. Burkhardt, et al., *Educational and Clinical Interventions* (pp. 151–177). Oxford, UK: Elsevier Science Ltd. DOI: 10.1016/S0270-4013(01)80012-X. 2

Dziobek, I., Fleck, S., Kalbe, E., Rogers, K., Hassenstab, J., Brand, M., Kessler, J., Woike, J. K., Wolf, O. T., and Convit, A. (2006). Introducing MASC: a movie for the assessment of social cognition. *Journal of Autism and Developmental Disorders*, 36(5), 623–636. DOI: 10.1007/s10803-006-0107-0. 46, 49, 50

Eder, M. S., Diaz, J. M. L., Madela, J. R. S., Magusara, M. U., and Sabellano, D. D. M. (2016). Fill me app: an interactive mobile game application for children with autism. *International Journal of Interactive Mobile Technologies* (iJIM), 10(3), 59–63. DOI: 10.3991/ijim.v10i3.5553. 62

Edmunds, S. R., Rozga, A., Li, Y., Karp, E. A., Ibanez, L. V., Rehg, J. M., and Stone, W. L. (2017). Brief report: using a point-of-view camera to measure eye gaze in young children with autism spectrum disorder during naturalistic social interactions: a pilot study. *Journal of Autism and Developmental Disorders*, 47(3), 898–904. DOI: 10.1007/s10803-016-3002-3. 111

Edrisinha, C., O'Reilly, M. F., Choi, H. Y., Sigafoos, J., and Lancioni, G. E. (2011). "Say Cheese": Teaching photography skills to adults with developmental disabilities. *Research in Developmental Disabilities*, 32(2), 636–642. DOI: 10.1016/j.ridd.2010.12.006. 42

Ertin, E., Stohs, N., Kumar, S., and Raij, A. (2011) AutoSense: unobtrusively wearable sensor suite for inferring the onset, causality, and consequences of stress in the field. In: *Proceedings of the 9th ACM Conference on Embedded Networked Sensor Systems*. New York: Association of Computing Machinery. 274–287. DOI: 10.1145/2070942.2070970.

Escobedo, L., Ibarra, C., Hernandez, J., Alvelais, M., and Tentori, M. (2014). Smart objects to support the discrimination training of children with autism. *Personal and Ubiquitous Computing*, 18(6), 1485–1497. DOI: 10.1007/s00779-013-0750-3. 124

Escobedo, L., Nguyen, D. H., Boyd, L., Hirano, S., Rangel, A., Garcia-Rosas, D., Tentori, M. and Hayes, G.R. (2012). MOSOCO: a mobile assistive tool to support children with autism practicing social skills in real-life situations. In *Proceedings of the 2012 ACM annual conference on Human Factors in Computing Systems* (pp. 2589–2598). ACM. DOI: 10.1145/2207676.2208649. 61, 64, 92

Eynon, A. (1997). Computer interaction: An update on the AVATAR program. *Communication*, Summer, 18, 1997. 87

Fabri, M. and Moore, D.J. (2005). The use of emotionally expressive avatars in collaborative virtual environments. In *Proceedings of Symposium on Empathic Interaction with Synthetic Characters. Symposium conducted at the Artificial Intelligence and Social Behaviour Convention* (AISB 2005), University of Hertfordshire, England. 89

Fabri, M., Elzouki, S .Y. A., and Moore, D. (2007). Emotionally expressive avatars for chatting, learning and therapeutic intervention. *Proceedings of the 12th International Conference on Human–Computer Interaction* (HCI International), 4552(3), 275–285. DOI: 10.1007/978-3-540-73110-8_29. 89

Farr, W., Yuill, N., and Raffle, H. (2010a). Social benefits of a tangible user interface for children with autistic spectrum conditions. *Autism*, 14(3), 237–252. DOI: 10.1177/1362361310363280. 123 , 124

Farr, W., Yuill, N., Harris, E., and Hinske, S. (2010b). In my own words: configuration of tangibles, object interaction and children with autism. In *Proceedings of the 9th International Conference on Interaction Design and Children* (pp. 30–38). ACM. DOI: 10.1145/1810543.1810548. 123

Fasoli, S. (2018). Transfer of upper limb robot-assisted training to daily activities after stroke: Considerations for patient-targeted home programs. *American Journal of Occupational Therapy*, 72(4_Supplement_1). DOI: 10.5014/ajot.2018.72S1-PO3029. 140

Fava, L. and Strauss, K. (2010). Multi-sensory rooms: Comparing effects of the Snoezelen and the stimulus preference environment on the behavior of adults with profound mental retardation. *Research in Developmental Disabilities*, 31(1), 160–171. DOI: 10.1016/j.ridd.2009.08.006. 95

Feil-Seifer, D. and Mataric, M. J. (2009). Toward socially assistive robotics for augmenting interventions for children with autism spectrum disorders. *Experimental Robotics*, 54, 201–210. DOI: 10.1007/978-3-642-00196-3_24. 12, 134, 135, 136, 141

Feil-Seifer, D. and Mataric, M. J. (2005). Defining socially assistive robotics. *Proceedings of the IEEE 9th International Conference on Rehabilitative Robotics* (ICORR 2005), June 28–July 1, Chicago, pp. 465–68. Piscataway, NJ: IEEE. DOI: 10.1109/ICORR.2005.1501143. 133

Feil-Seifer, D. and Mataric, M. J. (2008). B3 IA: A control architecture for autonomous robot-assisted behavior intervention for children with Autism Spectrum Disorders. In *Robot and Human Interactive Communication*, 2008. RO-MAN 2008. The 17th IEEE International Symposium on (pp. 328–333). IEEE. DOI: 10.1109/ROMAN.2008.4600687.

Feil-Seifer, D. and Mataric, M. J. (2011). Automated detection and classification of positive vs. negative robot interactions with children with autism using distance-based features. In *Proceedings of the ACM/IEEE International Conference on Human–Robot Interaction* (pp. 323–330). New York: ACM Press. DOI: 10.1145/1957656.1957785. 145

Ferrari, E., Robins, B., and Dautenhahn, K. (2009). Therapeutic and educational objectives in robot assisted play for children with autism. In *RO-MAN 2009-The 18th IEEE International Symposium on Robot and Human Interactive Communication* (pp. 108-114). IEEE. 139, 140

Finkelstein, S., Nickel, A., Barnes, T., and Suma, E. A. (2010). Astrojumper: motivating children with autism to exercise using a VR game. In *CHI'10 Extended Abstracts on Human Factors in Computing Systems* (pp. 4189–4194). ACM. DOI: 10.1145/1753846.1754124. 91

Finkenauer, C., Pollmann, M. M., Begeer, S., and Kerkhof, P. (2012). Brief report: Examining the link between autistic traits and compulsive internet use in a non-clinical sample. *Journal of Autism and Developmental Disorders*, 42(10), 2252–2256. DOI: 10.1007/ s10803-012-1465-4. 53

Fisicaro, D., Pozzi, F., Gelsomini, M., and Garzotto, F. (2019). Engaging persons with neuro-developmental disorder with a plush social robot. *Proceedings of ACM Human Robot Interaction* (HRI 2019). ACM. DOI: 10.1109/HRI.2019.8673107. 139, 140, 141

Fletcher-Watson, S., Petrou, A., Scott-Barrett, J., Dicks, P., Graham, C., O'Hare, A., Pain, H., and McConachie, H. (2016). A trial of an iPad™ intervention targeting social communication skills in children with autism. Autism, 20(7), 771-782. DOI: 10.1177/1362361315605624. 66

Flores, M., Musgrove, K., Renner, S., Hinton, V., Strozier, S., Franklin, S., and Hil, D. (2012). A comparison of communication using the Apple iPad and a picture-based system. *Augmentative and Alternative Communication*, 28(2), 74–84. DOI: 10.3109/07434618.2011.644579. 57

Foley, B. E. and Staples, A. H. (2003). Developing *Augmentative and Alternative Communication* (AAC) and literacy interventions in a supported employment setting. *Topics in Language Disorders*, 23(4), 325. DOI: 10.1097/00011363-200310000-00007.

Fombonne, E. (1999). The epidemiology of autism: A review. *Psychological Medicine*, 29, 769–786. DOI: 10.1017/S0033291799008508. 3

Fombonne, E. (2002). Epidemiological estimates and time trends in rates of autism. *Molecular Psychiatry*, 7, S4–S6. DOI: 10.1038/sj.mp.4001162. 3

Fong, T., Nourbakhsh, I., and Dautenhahn, K. (2003). A survey of socially interactive robots. *Robotic Autonomous Systems*, 42:143–66. DOI: 10.1016/S0921-8890(02)00372-X. 133

Foster, S. L. and and Cone, J. D. (1986). Design and use of direct observation. In A.R. Ciminero, K. Calhoun, and H. E. Adams (Eds.) *Handbook of Behavioral Assessment* (pp. 253–354). Wiley, New York. 114

Frauenberger, C., Good, J., and Keay-Bright, W. (2011). Designing technology for children with special needs: bridging perspectives through participatory design. *CoDesign*, 7(1), 1–28. DOI: 10.1080/15710882.2011.587013. 12

Frauenberger, C., Good, J., Alcorn, A., and Pain, H. (2012). Supporting the design contributions of children with autism spectrum conditions. In *Proceedings of the 11th International Conference on Interaction Design and Children* (pp. 134–143). ACM. DOI: 10.1145/2307096.2307112. 12

Freina, L., Busi, M., Canessa, A., Caponetto, I., and Ott, M. (2014). Learning to cope with street dangers: an interactive environment for intellectually impaired. *Edulearn14*, (July), 972–980.

Fridhi, A., Benzarti, F., Frihida, A., and Amiri, H. (2018). Application of virtual reality and augmented reality in psychiatry and neuropsychology, in particular in the case of autistic spectrum disorder (ASD). *Neurophysiology*, 1–7. DOI: 10.1007/s11062-018-9741-3. 90

Fuglerud, K.S. and Solheim, I. (2018). The use of social robots for supporting language training of children. In Transforming Our World Through Design, Diversity and Education: Proceedings of Universal Design and Higher Education in Transformation Congress 2018 (p. 401). IOS Press. 139

Fujimoto, I., Matsumoto, T., de Silva, P. R. S., Kobayashi, M., and Higashi, M. (2011). Mimicking and evaluating human motion to improve the imitation skill of children with autism through a robot. *International Journal of Social Robotics*, 3(4), 349–357. DOI: 10.1007/s12369-011-0116-9.

Gajecki, M., Berman, A. H., Sinadinovic, K., Rosendahl, I., and Andersson, C. (2014). Mobile phone brief intervention applications for risky alcohol use among university students: a randomized controlled study. *Addiction Science & Clinical Practice*, 9(1), 11. DOI: 10.1186/1940-0640-9-11. 66

Gal, E., Bauminger, N., Goren-Bar, D., Pianesi, F., Stock, O., Zancanaro, M., and Weiss, P. L. T. (2009). Enhancing social communication of children with high-functioning autism through a co-located interface. *AI and Society*, 24(1), 75–84. DOI: 10.1007/s00146-009- 0199-0. 77, 78

Ganz, J. B., Hong, E. R., and Goodwyn, F. D. (2013). Effectiveness of the PECS Phase III app and choice between the app and traditional PECS among preschoolers with ASD. *Research in Autism Spectrum Disorders*, 7(8), 973-983. DOI: 10.1016/j.rasd.2013.04.003. 58

Garzotto, F. and Bordogna, M. (2010). Paper-based multimedia interaction as learning tool for disabled children. In *Proceedings of the 9th International Conference on Interaction Design and Children* (pp. 79–88). ACM. DOI: 10.1145/1810543.1810553. 121

Garzotto, F. and Gelsomini, M. (2016). Integrating virtual worlds and mobile robots in game-based treatment for children with intellectual disability. In *Virtual Reality Enhanced Robotic Systems for Disability Rehabilitation* (pp. 69–84). IGI Global. DOI: 10.4018/978-1-4666-9740-9.ch005. 100, 135, 136, 137, 140

Garzotto, F. and Gelsomini, M. (2018). Magic Room: A smart space for children with neurodevelopmental disorder. *IEEE Pervasive Computing*, 17(1), 38–48. DOI: 10.1109/MPRV.2018.011591060. 87, 101, 102

Garzotto, F., Gelsomini, M., and Kinoe, Y. (2017a). *Puffy : a Mobile Inflatable Interactive Companion for Children with Neurodevelopmental Disorder* (section 2). DOI: 10.1007/978-3-319-67684-5_29. 140, 141, 142

Garzotto, F., Gelsomini, M., Clasadonte, F., Montesano, D., and Occhiuto, D. (2016a). Wearable immersive storytelling for disabled children. *Proceedings of the International Working Conference on Advanced Visual Interfaces - AVI'16*, 196–203. DOI: 10.1145/2909132.2909256.

Garzotto, F., Gelsomini, M., Matarazzo, V., Messina, N., and Occhiuto, D. (2017b). XOOM: An end-user development tool for web-based wearable immersive virtual tours. In

Lecture Notes in Computer Science (including subseries Lecture Notes in Artificial Intelligence and Lecture Notes in Bioinformatics), (Vol. 10360 LNCS, pp. 507–519). DOI: 10.1007/978-3-319-60131-1_36. 85, 90

Garzotto, F., Gelsomini, M., Occhiuto, D., and Matarazzo, V. (2017c). *Wearable Immersive Virtual Reality for Children with Disability: A Case Study*, 478–483. DOI: 10.1145/3078072.3084312.

Garzotto, F., Gelsomini, M., Pappalardo, A., Sanna, C., Stella, E., and Zanella, M. (2016b). Using brain signals in adaptive smart spaces for disabled children. In *Proceedings of the 2016 CHI Conference Extended Abstracts on Human Factors in Computing Systems* (pp. 1684–1690). ACM. DOI: 10.1145/2851581.2892533. 101

Garzotto, F., Gelsomini, M., Pappalardo, A., Sanna, C., Stella, E., and Zanella, M. (2016c). Monitoring and adaptation in smart spaces for disabled children. In *Proceedings of the International Working Conference on Advanced Visual Interfaces - AVI '16* (pp. 224–227). DOI: 10.1145/2909132.2909283.

Garzotto, F., Valoriani, M., Gelsomini, M., and Bartoli, L. (2014). Touchless motion-based interaction for therapy of autistic children. In *Virtual, Augmented Reality and Serious Games for Healthcare* 1 (pp. 471–494). Springer, Berlin, Heidelberg. DOI: 10.1007/978-3-642-54816-1_23. 91, 121, 122

Gay, V., Leijdekkers, P., Agcanas, J., Wong, F., and Wu, Q. (2013). CaptureMyEmotion: Helping autistic children understand their emotions using facial expression recognition and mobile technologies. In *Bled eConference*, (p. 10). 59

Gelsomini, M., (2018). Empowering interactive technologies for children with neuro-developmental disorders and their caregivers (http://hdl.handle.net/10589/137083). 65, 101, 143

Gelsomini, M., Cosentino, G., Spitale, M., Gianotti, M., Fisicaro, D., Leonardi, G., Riccardi, F., Piselli, A., Beccaluva, E., Bonadies, B., Di Terlizzi, L., Zinzone, M., Alberti, S., Rebourg, C., Carulli, M., Garzotto, F., Arquilla, V., Bisson, M., Del Curto, B., and Bordegoni, M. (2019). Magika, a multisensory environment for play, education and inclusion. In *Extended Abstracts of the 2019 CHI Conference on Human Factors in Computing Systems* (p. LBW0277). ACM. DOI: 10.1145/3290607.3312753.

Gelsomini, M., Garzotto, F., Matarazzo, V., Messina, N., and Occhiuto, D. (2017). Creating social stories as wearable hyper-immersive virtual reality experiences for children with neurodevelopmental disorders. *IDC 2017 - Proceedings of the 2017 ACM Conference on Interaction Design and Children*, (July), 431–437. DOI: 10.1145/3078072.3084305. 90, 135, 138

Gelsomini, M., Garzotto, F., Montesano, D., and Occhiuto, D. (2016). Wildcard: A wearable virtual reality storytelling tool for children with intellectual developmental disability. *Proceedings of the Annual International Conference of the IEEE Engineering in Medicine and Biology Society, EMBS*, 2016–October, 5188–5191. DOI: 10.1109/EMBC.2016.7591896. 85, 87, 90

Gelsomini, M., Leonardi, G., Degiorgi, M., Garzotto, F., Penati, S., Silvestri, J., Ramuzat, M., and Clasadonte, F. (2017). Puffy—an inflatable mobile interactive companion for children with neurodevelopmental disorders. In *Proceedings of the 2017 CHI Conference Extended Abstracts on Human Factors in Computing Systems - CHI EA '17* (pp. 2599–2606). DOI: 10.1145/3027063.3053245.

Gelsomini, M., Leonardi, G., and Garzotto, F. (2019). Embodied learning in immersive smart spaces. In P*roceedings of the 2019 CHI Conference on Human Factors in Computing Systems - CHI '19*. DOI: 10.1145/3313831.3376667.

Gelsomini, M., Rotondaro, A., Cosentino, G., Gianotti, M., Riccardi, F., and Garzotto, F. (2018, November). On the effects of a nomadic multisensory solution for children's playful learning. In *Proceedings of the 2018 ACM International Conference on Interactive Surfaces and Spaces* (pp. 189–201). ACM. DOI: 10.1145/3279778.3279790. 99, 100, 103

Gernsbacher, M. A., Dawson, M., and Goldsmith, H. H. (2005). Three reasons not to believe in an autism epidemic. *Current Directions in Psychological Science*, 14, 55–58. DOI: 10.1111/j.0963-7214.2005.00334.x. 4

Ghanouni, P., Jarus, T., Zwicker, J. G., Lucyshyn, J., Fenn, B., and Stokley, E. (2019a). Design elements during development of videogame programs for children with autism spectrum disorder: Stakeholders' viewpoints. *Games for Health Journal*, 9(2). DOI: 10.1089/g4h.2019.0070.

Ghanouni, P., Jarus, T., Zwicker, J. G., Lucyshyn, J., Mow, K., and Ledingham, A. (2019b). Social stories for children with autism spectrum disorder: Validating the cContent of a virtual reality program. *Journal of Autism and Developmental Disorders*, 49(2), 660-668. DOI: 10.1007/s10803-018-3737-0. 12, 89

Ghaziuddin, M., Tsai, L., and Ghaziuddin, N. (1992). Comorbidity of autistic disorder in children and adolescents. *European Journal of Child and Adolescent Psychiatry*, 1, 209–213. DOI: 10.1007/BF02094180. 3

Giannopulu, I. and Pradel, G. (2012). From child-robot interaction to child-robot-therapist interaction: A case study in autism. *Applied Bionics and Biomechanics*, 9(2), 173–179. DOI: 10.3233/JAD-2011-0042.

Gillan, N. (2019). Social Anxiety in Adult Autism. Ph.D. Thesis. University of Exeter. 139

Gillberg, C., Ehlers, S., Schaumann, H., G. Jakobsson, G., Dahlgren, S. O., Lindblom, R., Bagen-holm, A., Tjuus, T., and Blinder, E. (1990). Autism under age 3 years: A clinical study of 28 cases referred for autistic symptoms in infancy. *Journal of Child Psychology and Psychiatry*, 31:921, 934. DOI: 10.1111/j.1469-7610.1990.tb00834.x. 109

Giullian, N., Ricks, D., Atherton, A., Colton, M., Goodrich, M., and Brinton, B. (2010). Detailed requirements for robots in autism therapy. *Conference Proceedings - IEEE International Conference on Systems, Man and Cybernetics*, (October), 2595–2602. DOI: 10.1109/ICSMC.2010.5641908.

Giusti, L., Zancanaro, M., Gal, E., and Weiss, P. L. T. (2011). Dimensions of collaboration on a tabletop interface for children with autism spectrum disorder. In *Proceedings of the SIG-CHI Conference on Human Factors in Computing Systems* (pp. 3295–3304). ACM. DOI: 10.1145/1978942.1979431. 57, 72, 78

Glenwright, M. and Agbayewa, A. S. (2012). Older children and adolescents with high-func-tioning autism spectrum disorders can comprehend verbal irony in computer-mediated communication. *Research in Autism Spectrum Disorders*, 6(2), 628–638. DOI: 10.1016/j.rasd.2011.09.013. 37

Goan M., Fujii H., and Okada, M. (2006). Child-robot interaction mediated by building blocks: from field observations in a public space. *Artificial Life and Robotics*, 10(1):45–48. DOI: 10.1007/s10015-005-0375-3. 134

Golan, O. and Baron-Cohen, S. (2006). Systemizing empathy: Teaching adults with asperger syn-drome or high functioning autism to recognize complex emotions using interactive media, *Development and Psychopathology*, 18: 591–617. DOI: 10.1017/ S0954579406060305. 33, 45, 46

Golan, O., Ashwin, E., Granader, Y., McClintock, S., Day, K., Leggett, V., and Baron-Cohen, S. (2010). Enhancing emotion recognition in children with autism spectrum conditions: An intervention using animated vehicles with real emotional faces. *Journal of Autism and Developmental Disorders*, 40, 269–279. DOI: 10.1007/s10803-009-0862-9. 89

Golan, O., Baron-Cohen, S., Hill, J. J., and Golan, Y. (2007). The "reading the mind in the voice" test-revised: A study of complex emotion recognition in adults with and without autism spectrum conditions. *Journal of Autism and Developmental Disorders*, 37(6), 1096– 1106. DOI: 10.1007/s10803-006-0252-5.

Goldiamond, I. (1974). Toward a constructional approach to social problems: ethical and constitutional issues raised by applied behavior analysis. *Behaviorism*, 2(1), 1–84. DOI: 10.5210/bsi.v11i2.92. 32

Goldsmith, T. R. and LeBlanc, L. A. (2004). Use of technology in interventions for children with autism. *Journal of Early and Intensive Behavior Intervention*, 1(2), 166–178. DOI: 10.1037/h0100287. 12, 41

Goncales, J., Lima, J., Malheiros, P., and Costa, P. (2009). Realistic simulation of a Lego Mindstorms NXT based robot. In *2009 IEEE International Conference on Control Applications* (pp. 1242–1247). IEEE. DOI: 10.1109/CCA.2009.5280986.

Gonçalves, N., Rodrigues, J. L., Costa, S., and Soares, F. (2012). Automatic detection of stereotyped hand flapping movements: two different approaches. In *2012 IEEE RO-MAN: The 21st IEEE International Symposium on Robot and Human Interactive Communication* (pp. 392–397). IEEE. DOI: 10.1109/ROMAN.2012.6343784. 111

Goodrich, M. A. and Schultz, A. C. (2007). Human-robot interaction: a survey. *Foundations and Trends in Human-Computer Interaction*, 1(3), 203–275. DOI: 10.1561/1100000005. 133

Goodrich, M. A., Colton, M., Brinton, B., Fujiki, M., Atherton, J. A., Robinson, L., Ricks, D., Maxfield, M. H,m and Acerson, A. (2012). Incorporating a robot into an autism therapy team. *IEEE Intelligent Systems*, 27(2), 52–59. DOI: 10.1109/MIS.2012.40.

Goodwin, M. S., Groden, J., Velicer, W. F., Lipsitt, L. P., Baron, M. G., Hofmann, S. G., and Groden, G.(2006). Cardiovascular arousal in individuals with autism. *Focus on Autism and Other Developmental Disabilities*, 21, 100–123. DOI: 10.1177/10883576060210020101. 113, 115

Goodwin, M. S., Intille, S. S., Albinali, F., and Velicer, W. F. (2011). Automated detection of stereotypical motor movements. *Journal of Autism and Developmental Disorders*, 41, 770–782. DOI: 10.1007/s10803-010-1102-z. 114

Goodwin, M. S., Mazefsky, C. A., Ioannidis, S., Erdogmus, D., and Siegel, M. (2019). Predicting aggression to others in youth with autism using a wearable biosensor. *Autism Research*. DOI: 10.1002/aur.2151. 113, 116

Goodwin, M. S., Velicer, W. F., and Intille, S. S. (2008). Telemetric monitoring in the behavior sciences. *Behavioral Research Methods*, 40, 328–341. DOI: 10.3758/BRM.40.1.328. 107

Gotham, K., Pickles, A., and Lord, C. (2012). Trajectories of autism severity in children using standardized ADOS scores. *Pediatrics*, 130(5), e1278-e1284. DOI: 10.1542/peds.2011-3668. 108

Granpeesheh, D., Tarbox, J., Dixon, D. R., Peters, C. A., Thompson, K., and Kenzer, A. (2010). Evaluation of an eLearning tool for training behavioral therapists in academic knowledge of applied behavior analysis. *Research in Autism Spectrum Disorders*, 4(1), 11–17. DOI: 10.1016/j.rasd.2009.07.004. 39

Gray, C. (2003). *Social Stories 10.0.* Arlington, TX: Future Horizons. 43

Green, J., Gilchrist, A., Burton, D., and Cox, A. (2000). Social and psychiatric functioning in adolescents with Asperger syndrome compared with conduct disorder. *Journal of Autism and Developmental Disorders*, 30, 279–293. DOI: 10.1023/A:1005523232106. 8

Greffou, S., Bertone, A., Hahler, E., Hanssens, J., Mottron, L., and Faubert, J. (2012). Postural hypo-reactivity in autism is contingent on development and visual environment: A fully immersive virtual reality study. *Journal of Autism and Developmental Disorders*, 42, 961–970. DOI: 10.1007/s10803-011-1326-6. 91

Gresham, F. M. and MacMillan, D. L. (1997). Social competence and affective characteristics of students with mild disabilities. *Review of Educational Research*, 6, 377–415. DOI: 10.3102/00346543067004377. 9

Groden, J., Goodwin, M. S., Lipsitt, L. P., Hofmann, S. G., Baron, M. G., Groden, G., Velicer, W. F., and Plummer, B. (2005). Assessing cardiovascular responses to stressors in individuals with autism spectrum disorders. *Focus on Autism and Other Developmental Disorders*, 20, 244–252. DOI: 10.1177/10883576050200040601. 113

Grossard, C., Grynspan, O., Serret, S., Jouen, A. L., Bailly, K., and Cohen, D. (2017). Serious games to teach social interactions and emotions to individuals with autism spectrum disorders (ASD). *Computers and Education*, 113, 195–211. DOI: 10.1016/j.compedu.2017.05.002. 32

Großekathöfer, U., Manyakov, N. V., Mihajlović, V., Pandina, G., Skalkin, A., Ness, S., Bangerter, A., and Goodwin, M. S. (2017). Automated detection of stereotypical motor movements in autism spectrum disorder using recurrence quantification analysis. *Frontiers in Neuroinformatics*, 11, 9. DOI: 10.3389/fninf.2017.00009. 114

Grudin, J. (1994). Groupware and social dynamics: eight challenges for developers. *Communications of the ACM*, 37(1), 92–105. DOI: 10.1145/175222.175230. 75

Grynszpan, O., Martin, J.-C., and Nadel, J. (2008). Multimedia interfaces for users with high functioning autism: An empirical investigation. *International Journal of Human-Computer Studies*, 66(8), 628–639. DOI: 10.1016/j.ijhcs.2008.04.001. 45

Grynszpan, O., Weiss, P. L., Perez-Diaz, F., and Gal, E. (2013). Innovative technology based interventions for Autism Spectrum Disorders: A meta-analysis. *Autism*. DOI: 10.1177/1362361313476767. 12, 15, 25

Guazzaroni, G. and Pillai, A. S. (2019). Virtual reality (VR) for school children with autism spectrum disorder (ASD): A way of rethinking teaching and learning. In *Virtual and Augmented Reality in Mental Health Treatment* (pp. 141–158). IGI Global. DOI: 10.4018/978-1-5225-7168-1.ch009. 84

Gumtau, S., Newland, P., Creed, C., and Kunath, S. (2005). MEDIATE – a responsive environment designed for children with autism. *Digital World Conference 2005*, 8. Retrieved from http://eprints.port.ac.uk/6828/1/MEDIATE_dundee.pdf%0Ahttp://ewic.bcs.org/content/ConWebDoc/3805. DOI: 10.14236/ewic/AD2005.14.

Hagiwara, T. and Myles, B. S. (1999). A multimedia social story intervention: Teaching skills to children with autism. *Focus on Autism and Other Developmental Disabilities*, 14(2), 82–95. DOI: 10.1177/108835769901400203. 44

Hailpern, J., Harris, A., La Botz, R., Birman, B., and Karahalios, K. (2012). Designing visualizations to facilitate multisyllabic speech with children with autism and speech delays. In *Proceedings of the Designing Interactive Systems Conference* (pp. 126–135). ACM. DOI: 10.1145/2317956.2317977. 126

Hailpern, J., Karahalios, K., and Halle, J. (2009a). Creating a spoken impact: encouraging vocalization through audio visual feedback in children with ASD. In *Proceedings of the SIGCHI Conference on Human Factors in Computing Systems* (pp. 453–462). ACM. DOI: 10.1145/1518701.1518774. 126, 128

Hailpern, J., Karahalios, K., Halle, J., Dethorne, L., and Coletto, M. K. (2009b). A3: Hci coding guideline for research using video annotation to assess behavior of nonverbal subjects with computer-based intervention. *ACM Transactions on Accessible Computing* (TACCESS), 2(2), 8. DOI: 10.1145/1530064.1530066. 126

Hailpern, J., Karahalios, K., Halle, J., DeThorne, L., and Coletto., M. (2008). A3: a coding guideline for HCI+autism research using video annotation. In *Proceedings of the 10th International ACM SIGACCESS Conference on Computers and Accessibility* (Assets '08). ACM, New York, 11–18. DOI: 10.1145/1414471.1414476. 47

Halabi, O., Elseoud, S. A., Alja'am, J. M., Alpona, H., Al-Hemadi, M., and Al-Hassan, D. (2017). Immersive virtual reality in improving communication skills in children with autism. *International Journal of Interactive Mobile Technologies* (IJIM), 11(2), 146. DOI: 10.3991/ijim.v11i2.6555.

Han, Y., Fathi, A., Abowd, G. D., and Rehg, J. (2012). Automated detection of mutual eye contact and joint attention using a single wearable camera system. Abstract presented at the *International Meeting for Autism Research* (IMFAR). INSAR.

Hayes, G. R., Custodio, V. E., Haimson, O. L., Nguyen, K., Ringland, K. E., Ulgado, R. R., Waterhouse, A., and Weiner, R. (2015). Mobile video modeling for employment interviews for individuals with autism. *Journal of Vocational Rehabilitation*, 43(3), 275–287. DOI: 10.3233/JVR-150775. 44

Hayes, G. R., Gardere, L. M., Abowd, G. D., and Truong, K. N. (2008). CareLog: a selective archiving tool for behavior management in schools. *Proceeding of the Twenty-Sixth Annual CHI Conference on Human Factors in Computing Systems - CHI , 111'08* (pp. 685–694). DOI: 10.1145/1357054.1357164. 48, 49, 51

Hayes, G. R., Hirano, S., Marcu, G., Monibi, M., Nguyen, D. H., and Yeganyan, M. (2010). Interactive visual supports for children with autism. *Personal and Ubiquitous Computing*, 14(7), 663–680. DOI: 10.1007/s00779-010-0294-8. 44, 57, 70, 71, 111, 112

Hayes, G. R. and Hosaflook, S. W. (2013). HygieneHelper: promoting awareness and teaching life skills to youth with autism spectrum disorder. In *Proceedings of the 12th International Conference on Interaction Design and Children* (IDC '13), PP. 539–542. DOI: 10.1145/2485760.2485860. 62

Hayes, G. R., Kientz, J. A., Truong, K. N., White, D. R., Abowd, G. D., and Pering, T. (2004). Designing capture applications to support the education of children with autism. In *UbiComp 2004: Ubiquitous Computing* (pp. 161–178). Springer Berlin Heidelberg. DOI: 10.1007/978-3-540-30119-6_10. 47, 81

Hayes, G. R., Yeganyan, M. T., Brubaker, J. R., O'Neal, L., and Hosaflook, S. W. (2013). *Using Mobile Technologies to Support Students in Work Transition Programs. Twenty-First Century Skills for Students with Autis*m. Katharina Boser and Matthew Goodwin, Eds. Brooke's Publishing. In Press. 62

Healey, J. (2000). Future possibilities in electronic monitoring of physical activity. *Research Quarterly for Exercise and Sport*, 71, 137–145. DOI: 10.1080/02701367.2000.11082797. 107

Heathers, J. A., Gilchrist, K. H., Hegarty-Craver, M., Grego, S., and Goodwin, M. S. (2019). An analysis of stereotypical motor movements and cardiovascular coupling in individuals on the autism spectrum. *Biological Psychology*, 142, 90–99. DOI: 10.1016/j.biopsycho.2019.01.004. 113, 116

Hedley, D., Uljarević, M., Cameron, L., Halder, S., Richdale, A., and Dissanayake, C. (2017). Employment programmes and interventions targeting adults with autism spectrum disorder: A systematic review of the literature. *Autism*, 21(8), 929–941. DOI: 10.1177/1362361316661855. 8

Heffner, J. L., Vilardaga, R., Mercer, L. D., Kientz, J. A., and Bricker, J. B. (2015). Feature-level analysis of a novel smartphone application for smoking cessation. The American Journal of Drug and Alcohol Abuse, 41(1), 68-73. DOI: 10.3109/00952990.2014.977486. 66

Heimann, M., Nelson, K. E., Tjus, T., and Gillberg, C. (1995). Increasing reading and communication skills in children with autism through an interactive multimedia computer program. *Journal of Autism and Developmental Disorders*, 25(5), 459–480. DOI: 10.1007/BF02178294. 45

Heinrichs, M. and Domes, G. (2008). Neuropeptides and social behaviour: effects of oxytocin and vasopressin in humans. *Progress in Brain Research*, 170, 337–350. DOI: 10.1016/ S0079-6123(08)00428-7. 50

Hendricks, D. (2010). Employment and adults with autism spectrum disorders: Challenges and strategies for success. *Journal of Vocational Rehabilitation*, 32(2), 125–134. DOI: 10.3233/JVR-2010-0502. 8

Heni, N. and Hamam, H. (2016). Design of emotional educational system mobile games for autistic children. In *2016 2nd International Conference on Advanced Technologies for Signal and Image Processing* (ATSIP) (pp. 631–637). IEEE. DOI: 10.1109/ATSIP.2016.7523168. 59

Herrera, G., Alcantud, F., Jordan, R., Blanquer, A., Labajo, G., and De Pablo, C. (2008). Development of symbolic play through the use of virtual reality tools in children with autistic spectrum disorders. *Autism*, 12, 143–157. DOI: 10.1177/1362361307086657. 88

Hetzroni, O. E. and Tannous, J. (2004). Effects of a computer-based intervention program on the communicative functions of children with autism. *Journal of Autism and Developmental Disorders*, 34(2), 95–113. DOI: 10.1023/B:JADD.0000022602.40506.bf. 34

Higgins, K. and Boone, R. (1996). Creating individualized computer-assisted instruction for students with autism using multimedia authoring software. *Focus on Autism and Other Developmental Disabilities*, 11(2), 69–78. DOI: 10.1177/108835769601100202. 45

Hill, E., Berthoz, S., and Frith, U. (2004). Brief report: Cognitive processing of own emotions in individuals with autistic spectrum disorder and in their relatives. *Journal of Autism and Developmental Disorders*, 34, 229–235. DOI: 10.1023/B:JADD.0000022613.41399.14. 113

Hinckley, K. (2002). Input technologies and techniques. *The Human–Computer Interaction Handbook: Fundamentals, Evolving Technologies and Emerging Applications*, 151–168. DOI: 10.1201/b10368-12. 119

Hirano, S. H., Yeganyan, M. T., Marcu, G., Nguyen, D. H., Boyd, L. A., and Hayes, G.R. (2010). vSked: evaluation of a system to support classroom activities for children with autism.111 In *Proceedings of the 28th International Conference on Human Factors in Computing Systems - CHI '10 (*p. 1633). New York: ACM Press. DOI: 10.1145/1753326.1753569. 71, 78, 79

Hodges, S., Williams, L., Berry, E., Izadi, S., Srinivasan, J., Butler, A., Smyth, G., Kapur, N., and Wood, K. (2006). SenseCam: A retrospective memory aid. In *UbiComp 2006: Ubiquitous Computing* (pp. 177–193). Springer Berlin Heidelberg. DOI: 10.1007/11853565_11. 111

Hogg, J., Cavet, J., Lambe, L., and Smeddle, M. (2001). The use of "Snoezelen" as multisensory stimulation with people with intellectual disabilities: A review of the research. *Research in Developmental Disabilities*, 22(5), 353–372. DOI: 10.1016/S0891-4222(01)00077-4.

Holden, M. K. (2005). Virtual environments for motor rehabilitation. *Cyberpsychology & Behavior*, 8(3), 187-211. DOI: 10.1089/cpb.2005.8.187. 87

Hong, H., Kim, J. G., Abowd, G. D., and Arriaga, R. I. (2012). Designing a social network to support the independence of young adults with autism. In *Proceedings of the ACM 2012 Conference on Computer Supported Cooperative Work* (pp. 627–636). ACM. DOI: 10.1145/2145204.2145300. 101, 102

Hope, K. W. and Waterman, H. A. (2004). Using multi-sensory environments (MSEs) with people with dementia: Factors impeding their use as perceived by clinical staff. *Dementia*, 3(1), 45–68. DOI: 10.1177/1471301204039324.

Hopkins, I. M., Gower, M. W., Perez, T. A., Smith, D. A., Amthor, F. R., Wimsatt, F. C., and Biasini, F. J. (2011). Avatar assistant: Improving social skills in students with an ASD through a computer-based intervention. *Journal of Autism and Developmental Disorders*, 41, 1543–1555. DOI: 10.1007/s10803-011-1179-z. 33, 89

Hoshino, Y., Kaneko, M., Yashima, Y., Kumashiro, H., Volkmar, F. R., and and Cohen, D. J. (1987). Clinical features of autistic children with setback course in their infancy. *Japanese Journal of Psychiatry and Neurology*, 41:237, 245. DOI: 10.1111/j.1440-1819.1987.tb00407.x. 109

Hotz, G. A., Castelblanco, A., Lara, I. M., Weiss, A. D., Duncan, R., and Kuluz, J. W. (2006). Snoezelen: A controlled multi-sensory stimulation therapy for children recovering from severe brain injury. *Brain Injury*, 20(8), 879–888. DOI: 10.1080/02699050600832635. 95

Hourcade, J. P. (2008). Interaction design and children. *Found Trends Human Computer Interactions*, 1(4):277–392. DOI: 10.1561/1100000006.

Hourcade, J. P., Bullock-Rest, N. E., and Hansen, T. E. (2012). Multitouch tablet applications and activities to enhance the social skills of children with autism spectrum disorders. *Personal and Ubiquitous Computing*, 16(2), 157–168. DOI: 10.1007/s00779- 011-0383-3. 73

Hourcade, J. P., Williams, S. R., Miller, E. A., Huebner, K. E., and Liang, L. J. (2013). Evaluation of tablet apps to encourage social interaction in children with autism spectrum disorders. In *Proceedings of the 2013 ACM Annual Conference on Human Factors in Computing Systems* (3197–3206). ACM. DOI: 10.1145/2470654.2466438. 57, 61, 73

Howlin, P. and Goode, S. (1998). Outcome in adult life for people with autism and Asperger's syndrome. In: F. Volkmar (Ed.), *Autism and Pervasive Developmental Disorders* (pp. 209–241). Cambridge: University Press. DOI: 10.1177/1362361300004001005.

Howlin, P., Goode, S., Hutton, J., and Rutter, M. (2004). Adult outcome for children with autism. *Journal of Child Psychology and Psychiatry*, 45(2), 212–229. DOI: 10.1111/j.1469-7610.2004.00215.x. 8

Huijnen, C. A. G. J., Lexis, M. A. S., Jansens, R., and de Witte, L. P. (2016). Mapping robots to therapy and educational objectives for children with autism spectrum disorder. *Journal of Autism and Developmental Disorders*, 46(6), 2100–2114. DOI: 10.1007/s10803-016-2740-6. 134

Hulsegge, J. and Verheul, A. (1987). *Snoezelen: Another World: A Practical Book of Sensory Experience Environments for the Mentally Handicapped.* Rompa. 97

Huskens, B., Palmen, A., Van der Werff, M., Lourens, T., and Barakova, E. (2015). Improving collaborative play between children with autism spectrum disorders and their siblings: The effectiveness of a robot-mediated intervention based on Lego® Therapy. *Journal of Autism and Developmental Disorders*, 45(11), 3746–3755. DOI: 10.1007/s10803-014-2326-0.

Huskens, B., Verschuur, R., Gillesen, J., Didden, R., and Barakova, E. (2013). Promoting question-asking in school-aged children with autism spectrum disorders: Effectiveness of a robot intervention compared to a human-trainer intervention. *Developmental Neurorehabilitation*, 16(5), 345–356. DOI: 10.3109/17518423.2012.739212.

Iarocci, G. and McDonald, J. (2006). Sensory integration and the perceptual experience of persons with autism. *Journal of Autism and Developmental Disorders*, 36(1), 77–90. DOI: 10.1007/s10803-005-0044-3. 97

Intille, S. S., Larson, K., Beaudin, J. S., Nawyn, J., Tapia, E. M., and Kaushik, P. (2005). A living laboratory for the design and evaluation of ubiquitous computing technologies. In *CHI '05: Extended Abstracts on Human Factors in Computing systems* (pp. 1941–1944). New York: ACM Press. DOI: 10.1145/1056808.1057062. 107

Ip, H. H., Wong, S. W., Chan, D. F., Byrne, J., Li, C., Yuan, V. S., Lau, K. S. Y., and Wong, J. Y. (2018). Enhance emotional and social adaptation skills for children with autism spectrum disorder: A virtual reality enabled approach. *Computers and Education*, 117, 1–15. DOI: 10.1016/j.compedu.2017.09.010. 88

Ishii, H. and Ullmer, B. (1997). Tangible bits: toward seamless interfaces between people, bits and atoms. In *Proceedings of the ACM SIGCHI Conference on Human Factors in Computing Systems* (pp. 234–241). ACM. DOI: 10.1145/258549.258715. 122

James, W. (1950). *The Principles of Psychology*. New York: Dover. (Original work published 1890). 5

Jarrold, W., Mundy, P., Gwailtney, M., Hatt, N., McIntyre, N., Kim, K., Solomon, M., Novotny, S., and Swain, L. (2013). Social attention in a virtual public speaking task in higher functioning children with autism. *Autism Research*, Oct;6(5), 393–410. DOI: 10.1002/aur.1302. 89

Jazouli, M., Elhoufi, S., Majda, A., Zarghili, A., and Aalouane, R. (2016). Stereotypical motor movement recognition using microsoft kinect with artificial neural network. *World Academy of Science, Engineering and Technology International Journal of Computer and Electrrical Automation Control Information Engineering*, 10(7), 1270–1274. DOI: 10.1109/ICEMIS.2016.7745376. 111

Jensen, M., George, M. J., Russell, M. R., and Odgers, C. L. (2019). Young adolescents' digital technology use and mental health symptoms: Little evidence of longitudinal or daily linkages. *Clinical Psychological Science*, 7(6), 1416–1433. DOI: 10.1177/2167702619859336. 53

Jiang, X., Boyd, L. E., Chen, Y., and Hayes, G. R. (2016). ProCom: designing a mobile and wearable system to support proximity awareness for people with autism. In *Proceedings of the 2016 ACM International Joint Conference on Pervasive and Ubiquitous Computing: Adjunct* (pp. 93–96). ACM. DOI: 10.1145/2968219.2971445. 93

Jones, P., Wilcox, C., and Simon, J. (2016). Evidence-based instruction for students with autism spectrum disorder: Teachtown basics. In *Technology and the Treatment of Children with Autism Spectrum Disorder* (pp. 113–129). Springer, Cham. DOI: 10.1007/978-3-319-20872-5_10. 32

Jones, R. M., Southerland, A., Hamo, A., Carberry, C., Bridges, C., Nay, S., Stubbs, E., Komarow, E., Washington, C., Rehg, J., Lord, C., and Rozga, A. (2017). Increased eye contact during conversation compared to play in children with autism. *Journal of Autism and Developmental Disorders*, 47(3), 607–614. DOI: 10.1007/s10803-016-2981-4. 111

Jones, S. (2004). Augmentative and Alternative Communication: Management of severe communication disorders in children and adults. *Journal of Applied Research in Intellectual Disabilities*, 17(2), 133–134. DOI: 10.1111/j.1360-2322.2004.0182a.x.

Josman, N., Milika Ben-Chaim, H., Friedrich, S., and Weiss, P.L. (2008). Effectiveness of virtual reality for teaching street-crossing skills to children and adolescents with autism. *Interna-*

tional Journal on Disability and Human Development, 7(1), 49–56. http://www. degruyter. com/view/j/ijdhd. DOI: 10.1515/IJDHD.2008.7.1.49. 90

Kahn Jr, P. H., Kanda, T., Ishiguro, H., Freier, N. G., Severson, R. L., Gill, B. T., Ruckert, J. J., and Shen, S. (2012). "Robovie, you'll have to go into the closet now": Children's social and moral relationships with a humanoid robot. *Developmental Psychology*, 48(2), 303. DOI: 10.1037/a0027033. 137

Kaliouby, R. E., Picard, R., and Baron-Cohen, S. (2006). Affective computing and autism. *Annals of the New York Academy of Sciences*, 1093(1), 228–248. DOI: 10.1196/ annals.1382.016. 128

Kanero, J., Geçkin, V., Oranç, C., Mamus, E., Küntay, A. C., and Göksun, T. (2018). Social robots for early language learning: Current evidence and future directions. *Child Development Perspectives*. DOI: 10.1111/cdep.12277.

Kanner, L. (1943). Autistic disturbances of affective content. *Nervous Child*, 2, 217–250. 2

Kaplan, H., Clopton, M., Kaplan, M., Messbauer, L., and McPherson, K. (2006). Snoezelen multi-sensory environments: Task engagement and generalization. *Research in Developmental Disabilities*, 27(4), 443–455. DOI: 10.1016/j.ridd.2005.05.007. 95

Karanfiller, T., Göksu, H., and Yurtkan, K. (2017). A mobile application design for students who need special education. *Education and Science/Egitim ve Bilim*, 42(192). DOI: 10.15390/ EB.2017.7146. 62

Karray, F., Alemzadeh, M., Saleh, J. A., and Arab, M. N. (2008). Human-computer interaction: Overview on state of the art. *International Journal on Smart Sensing and Intelligent Systems*, 1(1). DOI: 10.21307/ijssis-2017-283. 17

Kaufman, D. R., Cronin, P., Rozenblit, L., Voccola, D., Horton, A., Shine, A., and Johnson, S. B. (2011). Facilitating the iterative design of informatics tools to advance the science of autism. *Studies in Health Technology and Informatics*, 169, 955. DOI: 10.3233/978-1-60750-806-9-955. 12

Kaur, M., Gifford, T., Marsh, K. L., and Bhat, A. (2013). Effect of robot–child interactions on bilateral coordination skills of typically developing children and a child with autism spectrum disorder: A preliminary study. *Journal of Motor Learning and Development*, 1(March 2016), 31–37. DOI: 10.1123/jmld.1.2.31.

Kaye, J. A., Maxwell, S. A., Mattek, N., Hayes, T. L., Dodge, H., Pavel, M., Jimison, H. B., Wild, K., Boise, L., and Zitzelberger, T. A. (2011). Intelligent systems for assessing aging changes: home-based, unobtrusive, and continuous assessment of aging. *Journals of Gerontology Series B: Psychological Sciences and Social Sciences*, 66(suppl_1), i180–i190. DOI: 10.1093/ geronb/gbq095. 107

Keay-Bright, W. and Howarth, I. (2012). Is simplicity the key to engagement for children on the autism spectrum? *Personal and Ubiquitous Computing*, 16(2), 129–141. DOI: 10.1007/s00779-011-0381-5. 123

Kenny, L., Hattersley, C., Molins, B., Buckley, C., Povey, C., and Pellicano, E. (2016). Which terms should be used to describe autism? Perspectives from the UK autism community. *Autism*, 20(4), 442–462. DOI: 10.1177/1362361315588200. viii

Kertz, S. J., Kelly, J. M., Stevens, K. T., Schrock, M., and Danitz, S. B. (2017). A review of free iPhone applications designed to target anxiety and worry. *Journal of Technology in Behavioral Science*, 2(2), 61-70. DOI: 10.1007/s41347-016-0006-y. 66

Keshav, N., Vahabzadeh, A., Abdus-Sabur, R., Huey, K., Salisbury, J., Liu, R., and Sahin, N. (2018). Longitudinal socio-emotional learning intervention for autism via smartglasses: Qualitative school teacher descriptions of practicality, usability, and efficacy in general and special education classroom settings. *Education Sciences*, 8(3), 107. DOI: 10.3390/educsci8030107. 93, 94

Kientz, J. A. (2012). Embedded capture and access: encouraging recording and reviewing of data in the caregiving domain. *Personal and Ubiquitous Computing*, 16(2), 209–221. DOI: 10.1007/s00779-011-0380-6. 35, 48, 56, 120, 121

Kientz, J. A. and Abowd, G. D. (2008). When the designer becomes the user: designing a system for therapists by becoming a therapist. In *CHI'08 Extended Abstracts on Human Factors in Computing Systems* (pp. 2071–2078). ACM. DOI: 10.1145/1358628.1358639. 120

Kientz, J. A., Arriaga, R. I., and Abowd, G. D. (2009). Baby Steps: evaluation of a system to support record-keeping for parents of young children. In *Proceedings of the SIGCHI Conference on Human Factors in Computing Systems* (pp. 1713–1722). ACM. DOI: 10.1145/1518701.1518965. 35, 49

Kientz, J. A., Hayes, G., Westeyn, T., Starner, T., and Abowd, G. D. (2007). Pervasive computing and autism: Assisting caregivers of children with special needs. *IEEE Pervasive Computing*, 6(1), 28–35. DOI: 10.1109/MPRV.2007.18. 12, 47, 48

Kientz, J. and Abowd, G. D. (2009). KidCam: Toward an effective technology for the capture of children's moments of interest. In *Pervasive Computing* (pp. 115–132). Springer Berlin Heidelberg. DOI: 10.1007/978-3-642-01516-8_9. 49, 56

Kientz, J., Boring, S., Abowd, G. D., and Hayes, G. (2005). Abaris: Evaluating automated capture applied to structured autism interventions. In *UbiComp 2005: Ubiquitous Computing* (pp. 323–339). Springer Berlin Heidelberg. DOI: 10.1007/11551201_19. 24, 48, 49, 120, 121, 126, 128, 129

Kientz, J., Hayes, G., Abowd, G. D., and Grinter, R. E. (2006). From the war room to the living room: decision support for home-based therapy teams. *Proceedings of the 2006 20th Anniversary Conference on Computer Supported Cooperative Work* (218–226). DOI: 10.1145/1180875.1180909. 24, 49, 120, 121, 126

Kijima, R., Shirakawa, K., Hirose, M., and Nihei, K. (1994). Virtual sand box: development of an application of virtual environments for clinical medicine. *Presence: Teleoperators and Virtual Environments*, 3, 45–59. DOI: 10.1162/pres.1994.3.1.45. 87

Kim, E. S., Berkovits, L. D., Bernier, E. P., Leyzberg, D., Shic, F., Paul, R., and Scassellati, B. (2013). Social robots as embedded reinforcers of social behavior in children with autism. *Journal of Autism and Developmental Disorders*, 43(5), 1038–1049. DOI: 10.1007/s10803-012-1645-2.

Kimball, J. W., Kinney, E. M., Taylor, B. A., and Stromer, R. (2004). Video enhanced activity schedules for children with autism: A promising package for teaching social skills. *Education and Treatment of Children*, 27(3). 44, 45

King, A. M., Brady, K. W., and Voreis, G. (2017). "It's a blessing and a curse": Perspectives on tablet use in children with autism spectrum disorder. *Autism and Developmental Language Impairments*, 2. DOI: 10.1177/2396941516683183. 57

King, A. M., Thomeczek, M., Voreis, G., and Scott, V. (2014). iPad® use in children and young adults with autism spectrum disorder: An observational study. *Child Language Teaching and Therapy*, 30(2), 159–173. DOI: 10.1177/0265659013510922. 55, 57

Kinney, E. M., Vedora, J., and Stromer, R. (2003). Computer-presented video models to teach generative spelling to a child with an autism spectrum disorder. *Journal of Positive Behavior Interventions*, 5(1), 22. Retrieved from http://www. ncbi.nlm.nih.gov/entrez/query.fcgi?db=-pubmed&cmd=Retrieve&dopt=Ab-stractPlus&list_uids=8576661618044703626related:ihNI7kxsBncJ. DOI: 10.1177/10983007030050010301. 43

Kinsella, B. G., Chow, S., and Kushki, A. (2017). Evaluating the usability of a wearable social skills training technology for children with autism spectrum disorder. *Frontiers in Robotics and AI*, 4, 31. DOI: 10.3389/frobt.2017.00031. 112

Kleeberger, V. and Mirenda, P. (2010). Teaching generalized imitation skills to a preschooler with autism using video modeling. *Journal of Positive Behavior Interventions*, 12, 116–127. DOI: 10.1177/1098300708329279. 42

Klin, A. and Volkmar, F. R. (2000). Treatment and intervention guidelines for individuals with Asperger syndrome. In A. Klin, F. R. Volkmar, and S. S. Sparrow (Eds.), *Asperger Syndrome* (pp. 340–366). New York: Guilford. 8, 134

Klin, A., Jones, W., Schultz, R., Volkmar, F., and Cohen, D. (2002). Visual fixation patterns during viewing of naturalistic social situations as predictors of social competence in individuals with autism. *Archives of General Psychiatry*, 59(9), 809. DOI: 10.1001/archpsyc.59.9.809. 128, 129

Klin, A., Lang, J., Cicchetti, D. V., and Volkmar, F. R. (2000). Interrater reliability of clinical diagnosis and DSM-IV criteria for autistic disorder: results of the DSM-IV autism field trial. *Journal of Autism Developmental Disorders*. 30(2):163–67. DOI: 10.1023/A:1005415823867.

Klin, A., Lin, D. J., Gorrindo, P., Ramsay, G., and Jones, W. (2009). Two-year-olds with autism orient to non-social contingencies rather than biological motion. *Nature*, 459(7244), 257–261. DOI: 10.1038/nature07868. 5

Klin, A., Volkmar, F. R., and Sparrow, S. S. (2005). *Asperger Syndrome*. New York: Guilford. DOI: 10.1002/9780470939345.ch4. 111

Kodak, T., Fisher, W. W., Clements, A., and Bouxsein, K. J. (2011). Effects of computer-assisted instruction on correct responding and procedural integrity during early intensive behavioral intervention. *Research in Autism Spectrum Disorders*, 5(1), 640–647. DOI: 10.1016/j.rasd.2010.07.011. 40

Koegel, L. K., Koegel, R. L., Harrower, J. K., and Carter, C. M. (1999). Pivotal response intervention I: Overview of approach. *Research and Practice for Persons with Severe Disabilities*, 24(3), 174–185. DOI: 10.2511/rpsd.24.3.174. 32

Koumpouros, Y. and Kafazis, T. (2019). Wearables and mobile technologies in Autism Spectrum Disorder interventions: A systematic literature review. *Research in Autism Spectrum Disorders*, 66, 101405. DOI: 10.1016/j.rasd.2019.05.005. 117

Kozima, H. and Zlatev, J. (2000). An epigenetic approach to human-robot communication. *Proceedings - IEEE International Workshop on Robot and Human Interactive Communication*, 346–351. DOI: 10.1109/ROMAN.2000.892521.

Kozima, H., Michalowski, M. P., and Nakagawa, C. (2009). Keepon: A playful robot for research, therapy, and entertainment. *International Journal of Social Robotics*, 1(1), 3–18. DOI: 10.1007/s12369-008-0009-8.

Kozima, H., Nakagawa, C., and Yasuda, Y. (2005). Interactive robots for communication-care: a case-study in autism therapy. *Proceedings of the 14th IEEE Int. Workshop Robot Hum. Interact. Commun.* (RO-MAN 2005), Aug. 13–15, Nashville, TN, (pp. 341–46). Piscataway, NJ: IEEE. DOI: 10.1109/ROMAN.2005.1513802. 135

Kozima, H., Nakagawa, C., and Yasuda, Y. (2007). Children-robot interaction: a pilot study in autism therapy. *Progress in Brain Research*, 164:385–400. DOI: 10.1016/S0079-6123(07)64021- 7. 135, 136

Kraleva, R. S. (2017). ChilDiBu–A mobile application for Bulgarian children with special educational needs. *International Journal on Advanced Science, Engineering and Information Technology* 7(6), 2085–2091. DOI: 10.18517/ijaseit.7.6.2922. 62

Kristin, S. and Solheim, I. (2018). The use of social robots for supporting language training of children. In *Transforming Our World Through Design, Diversity and Education: Proceedings of Universal Design and Higher Education in Transformation Congress 2018* (p. 401). IOS Press. DOI: 10.3233/978-1-61499-923-2-401.

Kumar, S., Nilsen, W. J., Abernethy, A., Atienza, A., Patrick, K., Pavel, M., Riley, W. T., Shar, A., Spring, B., Spruijt-Metz, D., Hedeker, D., Honavar, V., Kravitz, R, Lefebvre, R. C., Mohr, D. C., Murphy, S. A., Quinn, C., Shusterman, V., and Swendeman, D. (2013). Mobile health technology evaluation: The mHealth evidence workshop. *American Journal of Preventive Medicine*, 45, 228–236. DOI: 10.1016/j.amepre.2013.03.017. 113

Kushki, A., Drumm, E., Pla Mobarak, M., Tanel, N., Dupuis, A., Chau, T., and Anagnostou, E. (2013) Investigating the autonomic nervous system response to anxiety in children with autism spectrum disorders. *PLoS ONE*, 8(4): e59730. DOI: 10.1371/journal.pone.0059730. 113

Kwok, H. W. M., To, Y. F., and Sung, H. F. (2003). The application of a multisensory Snoezelen room for people with learning disabilities - Hong Kong experience. *Hong Kong Medical Journal*, 9(2), 122–126.

LaCava, P. G., Rankin, A., Mahlios, E., Cook, K., and Simpson, R. L. (2010). A single case design evaluation of a software and tutor intervention addressing emotion recognition and social interaction in four boys with ASD. *Autism*, 14(3), 161–178. DOI: 10.1177/1362361310362085. 33

LaCava, P., Golan, O., Baron-Cohen, S. and Myles, B.S. (2007) Using assistive technology to teach emotion recognition to students with asperger syndrome. *Remedial and Special Education*, 28: 174–181. DOI: 10.1177/07419325070280030601. 33

Lahav, O. and Mioduser, D. (2008). Construction of cognitive maps of unknown spaces using a multi-sensory virtual environment for people who are blind. *Computers in Human Behavior*, 24(3), 1139–1155. DOI: 10.1016/j.chb.2007.04.003. 95

Lahiri, U., Warren, Z., and Sarkar, N. (2011). Design of a Gaze-sensitive virtual social interactive system for children with autism. *IEEE Transactions on Neural Systems and Rehabilitation Engineering*, 19(4), 443–452. DOI: 10.1109/TNSRE.2011.2153874. 89

Lahm, E. A. (1996). Software that engages young children with disabilities: a study of design features. *Focus on Autism and Other Developmental Disabilities*, 11(2), 115–124. DOI: 10.1177/108835769601100207. 12

Lamash, L., Klinger, E., and Josman, N. (2017). Using a virtual supermarket to promote independent functioning among adolescents with autism spectrum disorder. In 2017 *International Conference on Virtual Rehabilitation* (ICVR) (pp. 1–7). IEEE. DOI: 10.1109/ICVR.2017.8007467. 90

Lang, R., O'Reilly, M., Healy, O., Rispoli, M., Lydon, H., Streusand, W., Davis, T., Sigafoos, J., Lancioni, G., Didden, R., and Giesbers, S. (2012). Sensory integration therapy for autism spectrum disorders: A systematic review. *Research in Autism Spectrum Disorders*, 6(3), 1004–1018. DOI: 10.1016/j.rasd.2012.01.006.

Lányi, C. S., Geiszt, Z., Károlyi, P., Tilinger, Á., and Magyar, V. (2006). Virtual reality in special needs early education. *The International Journal of Virtual Reality*, 5(4), 55–68.

Leaf, J. B., Leaf, R., McEachin, J., Taubman, M., Ala'i-Rosales, S., Ross, R. K., Smith, T., and Weiss, M. J. (2016). Applied behavior analysis is a science and, therefore, progressive. *Journal of Autism and Developmental Disorders*, 46(2), 720–731. DOI: 10.1007/s10803-015-2591-6. 32

Lee, J. H., Ku, J., Cho, W., Hahn, W. Y., Kim, I. Y., Lee, S.-M., Kang, U., Kim, D. Y., Yu, T., Wiederhold, B. K., Wiederhold, M. D., Kim, S. I. (2003). A virtual reality system for the assessment and rehabilitation of the activities of daily living. *Cyberpsychology and Behavior : The Impact of the Internet, Multimedia and Virtual Reality on Behavior and Society*, 6(4), 383–388. DOI: 10.1089/109493103322278763.

Lee, J., Takehashi, H., Nagai, C., Obinata, G., and Stefanov, D. (2012). Which robot features can stimulate better responses from children with autism in robot-assisted therapy? *International Journal of Advanced Robotic System*s, 9. DOI: 10.5772/51128.

Leekam, S. R., Nieto, C., Libby, S. J., Wing, L., and Gould, J. (2007). Describing the sensory abnormalities of children and adults with autism. *Journal of Autism and Developmental Disorders*, 37(5), 894–910. DOI: 10.1007/s10803-006-0218-7.

Leong, H. M., Carter, M., and Stephenson, J. (2015). Systematic review of sensory integration therapy for individuals with disabilities: Single case design studies. *Research in Developmental Disabilities*, 47, 334–351. DOI: 10.1016/j.ridd.2015.09.022.

Levine, T., Conradt, E., Goodwin, M. S., Sheinkopf, S., and Lester, B. (2014). Psychophysiologic arousal to social stress in autism spectrum disorders. In Patel, V. B., Preedy, V. R., and Martin C. (Eds.) *Comprehensive Guide to Autism*, 3 (pp. 1177–1194). Springer. DOI: 10.1007/978-1-4614-4788-7_66. 113

Lewis, L., Trushell, J., and Woods, P. (2005). Effects of ICT group work on interactions and social acceptance of a primary pupil with Asperger's Syndrome. *British Journal of Educational Technology*, 36(5), 739–755. DOI: 10.1111/j.1467-8535.2005.00504.x. 36

Lewis, V. and Boucher, J. (1988). Spontaneous, instructed and elicited play in relatively able autistic children. *British Journal of Developmental Psychology*, 6, 325–339. DOI: 10.1111/j.2044-835X.1988.tb01105.x. 5

Li, I., Dey, A. K., and Forlizzi, J. (2011). Understanding my data, myself: supporting self-reflection with ubicomp technologies. In *Proceedings of the 13th International Conference on Ubiquitous Computing* (pp. 405–414). ACM. DOI: 10.1145/2030112.2030166. 56

Li, K. H., Lou, S. J., Tsai, H. Y., and Shih, R. C. (2012). The effects of applying game-based learning to webcam motion sensor games for autistic students' sensory integration training. *Turkish Online Journal of Educational Technology-TOJET*, 11(4), 451–459. 122

Libin, A. V. and Libin, E. V. (2004). Person-robot interactions from the robopsychologists' point of view: The robotic psychology and robotherapy approach. *Proceedings of the IEEE*, 92(11), 1789–1803. DOI: 10.1109/JPROC.2004.835366. 138

Lin, H. Y., Perry, A., Cocchi, L., Roberts, J. A., Tseng, W. Y. I., Breakspear, M., and Gau, S. S. F. (2019). Development of frontoparietal connectivity predicts longitudinal symptom changes in young people with autism spectrum disorder. *Translational Psychiatry*, 9. DOI: 10.1038/s41398-019-0418-5. 8

Liu, C., Conn, K., Sarkar, N., and Stone, W. (2008a). Online affect detection and robot behavior adaptation for intervention of children with autism. *IEEE Transactions on Robotics*, 24, 883–896. DOI: 10.1109/TRO.2008.2001362. 138

Liu, C., Conn, K., Sarkar, N., and Stone, W. (2008b). Physiology-based affect recognition for computer-assisted intervention of children with autism spectrum disorder. *International Journal on Human-Computer Studies*, 66, 662–677. DOI: 10.1016/j.ijhsc.2008.04.003. 113

Liu, R., Salisbury, J. P., Vahabzadeh, A., and Sahin, N. T. (2017). Feasibility of an autism-focused augmented reality smartglasses system for social communication and behavioral coaching. *Frontiers in Pediatrics*, 5, 145. DOI: 10.3389/fped.2017.00145. 92

Loiacono, T., Trabucchi, M., Messina, N., Matarazzo, V., Garzotto, F., and Beccaluva, E. A. (2018). Social MatchUP-: A memory-like virtual reality game for the enhancement of social skills in children with neurodevelopmental disorders. In *Extended Abstracts of the 2018 CHI Conference on Human Factors in Computing Systems* (p. LBW619). ACM. DOI: 10.1145/3170427.3188525. 88, 89

Lopes, A. S. P., Araújo, J. V. M., Ferreira, M. P. V., and Ribeiro, J. E. M. (2015). A eficácia do Snoezelen na redução das estereotipias em adultos com deficiência intelectual: um estudo de caso da intervenção da terapia ocupacional em salas de estimulação multissensorial. *Revista de Terapia Ocupacional Da Universidade de São Paulo*, 26(2), 234. DOI: 10.11606/issn.2238-6149.v26i2p234-243.

Lorah, E. R., Parnell, A., Whitby, P. S., and Hantula, D. (2015). A systematic review of tablet computers and portable media players as speech generating devices for individuals with autism spectrum disorder. *Journal of Autism and Developmental Disorders*, 45(12), 3792–3804. DOI: 10.1007/s10803-014-2314-4. 57

Lord, C. and McGee, J. P. (2001). *Educating Children with Autism*. Washington, DC: National Academy Press. DOI: 10.17226/10017. 9

Lorenzo, G., Lledó, A., Pomares, J., and Roig, R. (2016). Design and application of an immersive virtual reality system to enhance emotional skills for children with autism spectrum disorders. *Computers and Education*, 98, 192–205. DOI: 10.1016/j.compedu.2016.03.018.

Lotan, M. (2006). Management of Rett syndrome in the controlled multisensory (Snoezelen) environment. A review with three case stories. *The Scientific World Journal*, 6, 791–807. DOI: 10.1100/tsw.2006.159.

Lotan, M. and Gold, C. (2009). Meta-analysis of the effectiveness of individual intervention in the controlled multisensory environment (Snoezelen®) for individuals with intellectual disability. *Journal of Intellectual and Developmental Disability*, 34(3), 207–215. DOI: 10.1080/13668250903080106.

Lotan, M. and Shapiro, M. (2005). Management of young children with Rett disorder in the controlled multi-sensory (Snoezelen) environment. *Brain and Development*, 27(SUPPL. 1), 88–94. DOI: 10.1016/j.braindev.2005.03.021.

Lotter, V. (1967). Epidemiology of autistic conditions in young children, II: Some characteristics of parents and their children. *Social Psychiatry*, 1, 163–173. DOI: 10.1007/ BF00578950. 3

Lovaas, O. I. (1987). Behavioral treatment and normal educational and intellectual functioning in young autistic children. *Journal of Consulting and Clinical Psychology*, 55(1), 3. DOI: 10.1037/0022-006X.55.1.3. 32

Lucas da Silva, M., Simões, C., Gonçalves, D., Guerreiro, T., Silva, H., and Botelho, F. (2011). TROCAS: communication skills development in children with autism spectrum disorders via ICT. In *Human-Computer Interaction–INTERACT 2011* (pp. 644–647). Springer Berlin Heidelberg. DOI: 10.1007/978-3-642-23768-3_103. 34

Lumbreras, M. A. M., de Lourdes, M. T. M., and Ariel, S. R. (2018). Aura: Augmented reality in mobile devices for the learning of children with ASD–Augmented reality in the learning of children with autism. In *Augmented Reality for Enhanced Learning Environments* (pp. 142–169). IGI Global. DOI: 10.4018/978-1-5225-5243-7.ch006. 94

Lydon, S., Healy, O., Reed, P., Mulhern, T., Hughes, B. M., and Goodwin, M. S. (2016). A systematic review of physiological reactivity to stimuli in autism. *Developmental Neurorehabilitation*, 19(6), 335–355. DOI: 10.3109/17518423.2014.971975.

Madsen, M., El Kaliouby, R., Eckhardt, M., Hoque, M. E., Goodwin, M. S., and Picard, R. (2009a). Lessons from participatory design with adolescents on the autism spectrum. In *CHI'09 Extended Abstracts on Human Factors in Computing Systems* (pp. 3835–3840). ACM. DOI: 10.1145/1520340.1520580. 12

Madsen, M., El-Kaliouby, R., Eckhardt, M., Goodwin, M. S., Hoque, M. E., and Picard, R.W. (2009b). Interactive social emotional toolkit (iSET). *Proceedings of the 2009 International Conference on Affective Computing and Intelligent Interaction*, September 10th–12th, Amsterdam, Netherlands. DOI: 10.1109/ACII.2009.5349531. 111, 128

Maione, L. and Mirenda, P. (2006). Effects of video modeling and video feedback on peer-directed social language skills of a child with autism. *Journal of Positive Behavior Interventions*, 8(2), 106–118. DOI: 10.1177/10983007060080020201. 43

Malinverni, L., Schaper, M. M., and Pares, N. (2016). An evaluation-driven design approach to develop learning environments based on full-body interaction. *Educational Technology Research and Development*, 64(6), 1337–1360. DOI: 10.1007/s11423-016-9468-z.

Mankoff, J., Hayes, G. R., and Kasnitz, D. (2010). Disability studies as a source of critical inquiry for the field of assistive technology. In *Proceedings of the 12th International ACM SIGACCESS Conference on Computers and Accessibility* (pp. 3–10). ACM. DOI: 10.1145/1878803.1878807. 10

Marco, E. J., Barett, L., Hinkley, N., and Hill, S. S. (2012). NIH Public access. *Pediatric Research*, 69, 1–14. DOI: 10.1203/PDR.0b013e3182130c54.Sensory.

Marco, J., Cerezo, E., and Baldassarri, S. (2013). Bringing tabletop technology to all: evaluating a tangible farm game with kindergarten and special needs children. *Personal and Ubiquitous Computing*, 17(8), 1577–1591. DOI: 10.1007/s00779-012-0522-5. 57

Marcotte, E. (2011). *Responsive Web Design*. Editions Eyrolles. 52

Marcu, G., Dey, A. K., and Kiesler, S. (2012). Parent-driven use of wearable cameras for autism support: a field study with families. In *Proceedings of the 2012 ACM Conference on Ubiquitous Computing* (pp. 401–410). ACM. DOI: 10.1145/2370216.2370277. 111

Marcu, G., Tassini, K., Carlson, Q., Goodwyn, J., Rivkin, G., Schaefer, K. J., and Kiesler, S. (2013). Why do they still use paper?: understanding data collection and use in Autism education. In *Proceedings of the SIGCHI Conference on Human Factors in Computing Systems* (pp. 3177–3186). ACM. DOI: 10.1145/2470654.2466436. 120, 130

Markopoulos, P. (2018). Mental Health Practitioners Perceptions' of Presence in a Virtual Reality Therapy Environment for Use for Children Diagnosed with Autism Spectrum Disorder. Ph.D. Thesis. University of New Orleans. 86, 89

Mars, A. E., Mauk, J. E., and Dowrick, P. W. (1998). Symptoms of pervasive developmental disorders as observed in prediagnostic home videos of infants and toddlers. *Journal of Pediatrics*, 132:500, 504. DOI: 10.1016/S0022-3476(98)70027-7. 109

Marti, P. (2010). Perceiving while being perceived. *International Journal of Design*, 4(2), 27–38.

Marti, P., Pollini, A., Rullo, A., and Shibata, T. (2005). Engaging with artificial pets. *Proceedings of the 2005 Annual Conference on European Association of Cognitive Ergonomics*, 99–106. Retrieved from http://dl.acm.org/citation.cfm?id=1124666.1124680. 138

Maskey, M., McConachie, H., Rodgers, J., Grahame, V., Maxwell, J., Tavernor, L., and Parr, J. R. (2019a). An intervention for fears and phobias in young people with autism spectrum disorders using flat screen computer-delivered virtual reality and cognitive behaviour therapy. *Research in Autism Spectrum Disorders*, 59, 58–67. DOI: 10.1016/j.rasd.2018.11.005. 90

Maskey, M., Rodgers, J., Grahame, V., Glod, M., Honey, E., Kinnear, J., Labus, M., Milne, J., Minos, D., McConachie, H., and Parr, J. R. (2019b). A randomised controlled feasibility trial of immersive virtual reality treatment with cognitive behaviour therapy for specific phobias in young people with autism spectrum disorder. *Journal of Autism and Developmental Disorders*, 1–16. DOI: 10.1007/s10803-018-3861-x. 90

Masuch, M. and Liszio, S. (2017). Participatory design of virtual reality applications for children under medical treatment. 16th *Interaction Design and Children Conference* (IDC 2017). Workshop "Analyzing Children's Contributions and Experiences in Co-design Activities."

Matson, J. L. and Nebel-Schwalm, M. (2007). Assessing challenging behaviors in children with autism spectrum disorders: A review. *Research in Developmental Disabilities*, 28, 567–578. DOI: 10.1016/j.ridd.2006.08.001. 114

Mazurek, M. O. (2013). Social media use among adults with autism spectrum disorders. *Computers in Human Behavior*, 29(4), 1709–1714. DOI: 10.1016/j.chb.2013.02.004. 36

Mazurek, M. O., Shattuck, P. T., Wagner, M., and Cooper, B. P. (2012). Prevalence and correlates of screen-based media use among youths with autism spectrum disorders. *Journal of Autism and Developmental Disorders*, 42(8), 1757–1767. DOI: 10.1007/s10803- 011-1413-8. 1

Mazurek, M. O., Engelhardt, C. R., and Clark, K. E. (2015). Video games from the perspective of adults with autism spectrum disorder. *Computers in Human Behavior*, 51, 122-130. DOI: 10.1016/j.chb.2015.04.062. 38

Mazzei, D., Billeci, L., Armato, A., Lazzeri, N., Cisternino, A., Pioggia, G., Igliozzi, R., Muratori, F., Ahluwalia, A., and De Rossi, D. (2010). The FACE of autism. *Proceedings - IEEE International Workshop on Robot and Human Interactive Communication*, (October), 791–796. DOI: 10.1109/ROMAN.2010.5598683. 138

Mazzei, D., Lazzeri, N., Billeci, L., Igliozzi, R., Mancini, A., Ahluwalia, A., Muratori, F., and de Rossi, D. (2011). Development and evaluation of a social robot platform for therapy in autism. *33rd Annual Conference of the IEEE EMBS*, Boston, MA, August 30th-September 3rd. DOI: 10.1109/IEMBS.2011.6091119. 138

Mazzei, D., Greco, A., Lazzeri, N., Zaraki, A., Lanata, A., Igliozzi, R., Mancini, A., Stoppa, F., Scilingo, E. P., Murtori, F., and De Rossi, D. (2012). Robotic social therapy on children with autism: Preliminary evaluation through multi-parametric analysis. *IEEE Conference on Social Computing*, 955–960. DOI: 10.1109/SocialCom-PASSAT.2012.99. 138

McComas, J., Pivik, J., and Laflamme, M. (1998). Current uses of virtual reality for children with disabilities. *Studies in Health Technology and Informatics*, 58(August), 161–169. DOI: 10.3233/978-1-60750-902-8-161.

McCoy, K. and Hermansen, E. (2007). Video modeling for individuals with autism: A review of model types and effects. *Education and Treatment of Children*, 30(4), 183–213. DOI: 10.1353/etc.2007.0029. 40

Mcduff, D., Hurter, C., and Gonzalez-Franco, M. (2017). Pulse and vital sign measurement in mixed reality using a HoloLens. In *Proceedings of the 23rd ACM Symposium on Virtual Reality Software and Technology* (p. 34). ACM. DOI: 10.1145/3139131.3139134. 92

McMahon, D., Cihak, D. F., and Wright, R. (2015). Augmented reality as a navigation tool to employment opportunities for postsecondary education students with intellectual disabilities and autism. *Journal of Research on Technology in Education*, 47(3), 157–172. DOI: 10.1080/15391523.2015.1047698. 94

McNaughton, D. and Chapple, D. (2013). AAC and communication in the workplace. *Perspectives on Augmentative and Alternative Communication*, 22(1), 30–36. DOI: 10.1044/aac22.1.30.

Mechling, L. C., Gast, D. L., and Seid, N. H. (2009). Using a personal digital assistant to increase independent task completion by students with autism spectrum disorder. *Journal of Autism and Developmental Disorders*, 39(10), 1420–1434. DOI: 10.1007/s10803-009- 0761-0. 62

Mehl, M. R., Pennebaker, J. W., Crow, D. M., Dabbs, J., and Price, J. H. (2001). The Electronically Activated Recorder (EAR): A device for sampling naturalistic daily activities and conversations. *Behavior Research Methods, Instruments, and Computers*, 33, 517–523. DOI: 10.3758/BF03195410. 107

Meltzoff, A. N., Brooks, R., Shon, A. P., and Rao, R. P. N. (2010). "Social" robots are psychological agents for infants: A test of gaze following. *Neural Networks*, 23(8–9), 966–972. DOI: 10.1016/j.neunet.2010.09.005.

Mesa-Gresa, P., Gil-Gómez, H., Lozano-Quilis, J. A., and Gil-Gómez, J. A. (2018). Effectiveness of virtual reality for children and adolescents with autism spectrum disorder: an evidence-based systematic review. *Sensors*, 18(8), 2486. DOI: /10.3390/s18082486. 87

Michaud, F., Salter, T., Duquette, A., Mercier, H., Lauria, M., Larouche, H., and Larose, F. (2007). Mobile robots engaging children in learning. *Canadian Medical and Biologiccal Engineering Conference*. Retrieved from http://introlab.3it.usherbrooke.ca/papers/FIC-CDAT2007e.pdf.

Michaud, F. and Théberge-Turmel, C. (2002). Mobile robotic toys and autism. In *Socially Intelligent Agents* (pp. 125–132). Springer US. DOI: 10.1007/0-306-47373-9_15. 135

Michaud, F., Duquette, A., and Nadeau, I. (2003). Characteristics of mobile robotic toys for children with pervasive developmental disorders. *SMC'03 Conference Proceedings. 2003 IEEE International Conference on Systems, Man and Cybernetics. Conference Theme - System Security and Assurance* (Cat. No.03CH37483), 3, 2938–2943. DOI: 10.1109/ICSMC.2003.1244338. 138

Michelle, R., Kandalaft, M. R., Didehbani,, N., Krawczyk, D. C., Allen, T. T., and Chapman, S. B. (2013). Virtual reality social cognition training for young adults with high-functioning autism. *Journal of Autism and Developmental Disorders*, 43, 34–44. DOI: 10.1007/ s10803-012-1544-6. 88

Milgram, P., Takemura, H., Utsumi, A., and Kishino, F. (1995). Augmented reality: A class of displays on the reality-virtuality continuum. In *Telemanipulator and Telepresence Technologies*, 2351, pp. 282–293). International Society for Optics and Photonics. DOI: 10.1117/12.197321. 84

Millen, L., Cobb, S., and Patel, H. (2011). Participatory design approach with children with autism. *International Journal on Disability and Human Development*, 10(4), 289–294. DOI: 10.1515/IJDHD.2011.048. 12

Milne, R., Clare, I. C. H., and Bull, R. (2002). Interrogative suggestibility among witnesses with mild intellectual disabilities: The use of an adaptation of the GSS. *Journal of Applied Research in Intellectual Disabilities*, 15(1), 8–17. DOI: 10.1046/j.1360-2322.2002.00102.x.

Milton, D. (2018). *A Critique of the Use of Applied Behavioural Analysis (ABA): On Behalf of the Neurodiversity Manifesto Steering Group*. University of Kent (https://kar.kent.ac.uk/69268/). 32

Milton, D. E. (2012). On the ontological status of autism: the 'double empathy problem'. *Disability and Society*, 27(6), 883–887. DOI: 10.1080/09687599.2012.710008. 10

Mineo, B. A., Ziegler, W., Gill, S. and Salkin, D. (2009). Engagement with electronic screen media among students with autism spectrum disorders. *Journal of Autism and Developmental Disorders*, 39, 172–187. DOI: 10.1007/s10803-008-0616-0. 87

Mirenda, P. (2001). Autism, augmentative communication, and assistive technology what do we really know? *Focus on Autism and Other Developmental Disabilities*, 16(3), 141–151. DOI: 10.1177/108835760101600302. 12

Mitchell, P., Parsons, S. and Leonard, A. (2007). Using virtual environments for teaching social understanding to adolescents with autistic spectrum disorders. *Journal of Autism and Developmental Disorders*, 37, 589–600. DOI: 10.1007/s10803-006-0189-8. 88

Mohamed, A. O., Courboulay, V., Sehaba, K., and Ménard, M. (2006). Attention analysis in interactive software for children with autism. In *Proceedings of the 8th International ACM SIGACCESS Conference on Computers and Accessibility* (pp. 133–140). ACM. DOI: 10.1145/1168987.1169011. 33, 127

Moir, L. (2010). Evaluating the effectiveness of different environments on the learning of switching skills in children with severe and profound multiple disabilities. *British Journal of Occupational Therapy*, 73(10), 446–456. DOI: 10.4276/030802210X12865330218186.

Mokashi, S., Yarosh, S., and Abowd, G. D. (2013). Exploration of videochat for children with autism. In *Proceedings of the 12th International Conference on Interaction Design and Children* (pp. 320–323). DOI: 10.1145/2485760.2485839. 31

Moore, D., Cheng, Y., McGrath, P., and Powell, N. J. (2005). Collaborative virtual environment technology for people with autism. *Focus on Autism and Other Developmental Disorders*, 20, 231–243. DOI: 10.1177/10883576050200040501. 89

Moore, M. and Calvert, S. (2000). Brief report: Vocabulary acquisition for children with autism: Teacher or computer instruction. *Journal of Autism and Developmental Disorders*, 30(4), 359–362. DOI: 10.1023/A:1005535602064. 39

Mora-Guiard, J., Crowell, C., Pares, N., and Heaton, P. (2016). Lands of fog: Helping children with autism in social interaction through a full-body interactive experience. In *Proceedings of the the 15th International Conference on Interaction Design and Children* (pp. 262–274). ACM. DOI: 10.1145/2930674.2930695. 99

Morris, M.R., Huang, A., Paepcke, A., and Winograd, T. (2006) Cooperative gestures: Multi-user gestural interactions for co-located groupware. In *Proceedings of CHI 2006*, 1201–1210. DOI: 10.1145/1124772.1124952. 69

Morris, R. R., Kirschbaum, C. R., and Picard, R. W. (2010). Broadening accessibility through special interests: a new approach for software customization. In *Proceedings of the 12th International ACM SIGACCESS Conference on Computers and Accessibility* (pp. 171–178). ACM. DOI: 10.1145/1878803.1878834. 11

Mottron, L., Dawson, M., Soulieres, I., Hubert, B., and Burack, J. (2006). Enhanced perceptual functioning in autism: An update, and eight principles of autistic perception. *Journal of Autism and Developmental Disorders*, 36, 27–43. DOI: 10.1007/s10803-005-0040-7. 3

Mount, H. and Cavet, J. (1995). Multi-sensory environments: an exploration of their potential for young people with profound and multiple learning difficulties. *British Journal of Special Education*, 22(2), 52–55. DOI: 10.1111/j.1467-8578.1995.tb01322.x.

Mower, E., Black, M. P., Flores, E., Williams, M., and Narayanan, S. (2011). Rachel: Design of an emotionally targeted interactive agent for children with autism. In *Multimedia and Expo (ICME), 2011 IEEE International Conference on* (pp. 1–6). IEEE. DOI: 10.1109/ICME.2011.6011990. 89

Mun, K. H., Kwon, J. Y., Lee, B. H., and Jung, J. S. (2014). Design developing an early model of cat robot for the use of early treatment of children with autism spectrum disorder (ASD). *International Journal of Control and Automation*, 7(11), 59–74. DOI: 10.14257/ijca.2014.7.11.07.

Mundy, P. and Sigman, M. (1989). Specifying the nature of the social impairment in autism. In G. Dawson (Ed.) *Autism: New Perspectives on Diagnosis, Nature, and Treatment* (p. 3–21). New York: Guilford Publications, Inc. 5

Mundy, P. and Neal, A.R. (2001). Neural plasticity, joint attention, and a transactional social-orienting model of autism. In L.M. Glidden (Ed.), *International Review of Research in*

Mental Retardation: Autism (vol. 23, pp. 139–168). San Diego, CA: Academic Press. DOI: 10.1016/S0074-7750(00)80009-9. 135

Mundy, P., Sigman, M., and Kasari, C. (1990). A longitudinal study of joint attention and language development in autistic children. *Journal of Autism and Developmental Disorders*, 20, 115–128. DOI: 10.1007/BF02206861. 135

Nahum-Shani, I., Smith, S. N., Spring, B. J., Collins, L. M., Witkiewitz, K., Tewari, A., and Murphy, S. A. (2018). Just-in-time adaptive interventions (jitais) in mobile health: Key components and design principles for ongoing health behavior support. *Annals of Behavioral Medicine*, 52(6) pp. 446–462. DOI: 10.1007/s12160-016-9830-8. 113

Narayanan, S. and Georgiou, P.G. (2013). Signal processing: Deriving human behavioral informatics from speech and language. *Proceedings of the IEEE, 2013*, 101, (5), 1203–1233. DOI: 10.1109/JPROC.2012.2236291. 31, 47

Navedo, J., Espiritu-Santo, A., and Ahmed, S. (2019). Strength-based ICT design supporting individuals with autism. In *The 21st International ACM SIGACCESS Conference on Computers and Accessibility* (pp. 560–562). DOI: 10.1145/3308561.3354637. 12

Nazneen, N., Rozga, A., Romero, M., Findley, A. J., Call, N. A., Abowd, G. D., and Arriaga, R. I. (2011). Supporting parents for in-home capture of problem behaviors of children with developmental disabilities. *Personal and Ubiquitous Computing*, 1–15–15. DOI: 10.1007/s00779-011-0385-1. 49

Neale, H. R., Kerr, S. J., Cobb, S. V. G., and Leonard, A. (2002). Exploring the role of virtual environments in the special needs classroom. *Proceedings of the 4th International Conference on Disability, Virtual Reality and Associate Technologies*, Veszprém, Hungary: 2002, ICD-VRAT/University of Reading, UK. 88

Newbutt, N. and Cobb, S. (2018). Towards a framework for implementation of virtual reality technologies in schools for autistic pupils. 255-258. Paper presented at *12th International Conference on Disability, Virtual Reality and Associated Technologies in Collaboration with Interactive Technologies and Games* (ITAG). 87

Newbutt, N., Sung, C., Kuo, H.-J., Leahy, M. J., Lin, C.-C., and Tong, B. (2016). Brief report: A pilot study of the use of a virtual reality headset in autism populations. *Journal of Autism and Developmental Disorders*, 46(9), 3166–3176. DOI: 10.1007/s10803-016-2830-5.

Nikopoulos, C. K., Canavan, C., and Nikopoulou-Smyrni, P. (2008). Generalized effects of video modeling on establishing instructional stimulus control in children with autism: Results of a preliminary study. *Journal of Positive Behavior Interventions*, 11(4), 198–207. DOI: 10.1177/1098300708325263. 42, 43

Niku, S. B. (2001). *Introduction to Robotics: Analysis, Systems, Applications* (Vol. 7). Englewood Cliffs, NJ: Prentice Hall. 133

Oberleitner, R., Ball, J., Gillette, D., Naseef, R., and Stamm, B. H. (2006). Technologies to lessen the distress of autism. *Journal of Aggression, Maltreatment and Trauma*, 12(1-2), 221–242. DOI: 10.1300/J146v12n01_12. 12

Ochs, E. and Solomon, O. (2010). Autistic sociality. *Ethos*, 38(1), 69–92. DOI: 10.1111/j.1548-1352.2009.01082.x. 138

Ochs, E., Kremer-Sadlik, T., Sirota, K. G., and Solomon, O. (2004). Autism and the social world: An anthropological perspective. *Discourse Studies*, 6(2), 147–183. DOI: 10.1177/1461445604041766. 138

Ochs, E., Kremer-Sadlik, T., Solomon, O., and Sirota, K. G. (2001). Inclusion as social practice: Views of children with autism. *Social Development*, 10(3), 399–419. DOI: 10.1111/1467-9507.00172. 138

Odom, S. L. and Diamond, K. E. (1998). Inclusion of young children with special needs in early childhood education: The research base. *Early Childhood Research Quarterly*, 13(1), 3–25. DOI: 10.1016/S0885-2006(99)80023-4. 10

Oller, D. K., Niyogi, P., Gray, S., Richards, J. A., Gilkerson, J., Xu, D., et al. (2010). Automated vocal analysis of naturalistic recordings from children with autism, language delay, and typical development. *Proceedings of the National Academy of Sciences of the United States of America*. DOI: 10.1073/pnas.1003882107. 112

Olley, G. J. (2005). Curriculum and classroom structure. In F. R. Volkmar, R. Paul, A. Klin, and D. Cohen (Eds.). *Handbook of Autism and Pervasive Developmental Disorders* (pp. 863–881). New Jersey: John Wiley and Sons. DOI: 10.1002/9780470939352.ch7. 9

Osterling, J. A., Dawson, G., and Munson, J. A. (2002). Early recognition of 1-year-old infants with autism spectrum disorder versus mental retardation. *Development and Psychopathology*, 14, 239–251. DOI: 10.1017/S0954579402002031. 5

Osterling, J. and Dawson, G. (1994). Early recognition of children with autism: A study of the first birthday home videotapes. *Journal of Autism and Developmental Disorders*, 24:247, 257. DOI: 10.1007/BF02172225. 109

Pagliano, P. (n.d.). *Using a Multisensory Environment- A Practical Guide for Teachers - Libro in lingua inglese*, Taylor and Francis Ltd., IBS. (n.d.).

Panyan, M. V. (1984). Computer technology for autistic students. *Journal of Autism and Developmental Disorders*, 14(4), 375-382. DOI: 10.1007/BF02409828. 11

Parchomiuk, M. (2019). Sexuality of persons with autistic spectrum disorders (ASD). *Sexuality and Disability*, 37(2), 259–274. DOI: 10.1007/s11195-018-9534-z. 8

Parenteau, R.E. (2011). Comparing the acquisition rates of stimulus presentation in discrete trials: table-top vs. scan-board. *Applied Behavioral Analysis Master's Theses*. Paper 57. http://hdl.handle.net/2047/d20001036. 70

Parés, N., Carreras, A., Durany, J., Ferrer, J., Freixa, P., Gómez, D., Kruglanski, O., Parés, R., Ribas, J. I., Soler, M., and Sanjurjo, A. (2004). MEDIATE: An interactive multisensory environment for children with severe autism and no verbal communication. In *Proceedings of the Third International Workshop on Virtual Rehabilitation*. 81, 98, 99

Parés, N., Masri, P., Van Wolferen, G., and Creed, C. (2005a). Achieving dialogue with children with severe autism in an adaptive multisensory interaction: The "MEDIATE" project. *IEEE Transactions on Visualization and Computer Graphics*, 11(6), 734–742. DOI: 10.1109/TVCG.2005.88. 81

Parés, N., Carreras, A., Durany, J.,Ferrer, J.,Freixa, P., Gómez, D., Kruglanski, O., Roc Paréz, J., Ribas,J. I., Soler, M., and Sanjurjo, À. (2005b). Promotion of creative activity in children with severe autism through visuals in an interactive multisensory environment. *Proceeding of the 2005 Conference on Interaction Design and Children*, IDC '05, (April 2016), 110–116. DOI: 10.1145/1109540.1109555. 98

Parette, P. and Scherer, M. (2004). Assistive technology use and stigma. *Education and Training in Developmental Disabilities*, 39(3), 217–226.

Parham, L. D., Cohn, E. S., Spitzer, S., Koomar, J. A., Miller, L. J., Burke, J. P.,Brette-Green, B., Mailloux, Z., May-Benson, T. A., Smith Roley, S., Schaaf, R. C., Schoen, S. A., and Summers, C. A. (2007). Fidelity in sensory integration intervention research. *American Journal of Occupational Therapy*, 61(2), 216–227. DOI: 10.5014/ajot.61.2.216. 96

Parry, C. (2017). An exploration of the potential therapeutic applications of the Amazon Echo and Dash devices in supporting students with autism dpectrum fisorders. Doctoral dissertation, Cardiff Metropolitan University. 126

Parsons, S. (2016). Authenticity in virtual reality for assessment and intervention in autism: A conceptual review. *Educational Research Review*, 19, 138–157. DOI: 10.1016/j.edurev.2016.08.001.

Parsons, S. and Cobb, S. (2011). State-of-the-art of virtual reality technologies for children on the autism spectrum. *European Journal of Special Needs Education*, 26(March 2015), 355–366. DOI: 10.1080/08856257.2011.593831. 103

Parsons, S. and Mitchell, P. (2002). The potential of virtual reality in social skills training for people with autistic spectrum disorders. *Journal of Intellectual Disability Research*, 46(5), 430–443. DOI: 10.1046/j.1365-2788.2002.00425.x. 12, 85

Parsons, S., Leonard, A., and Mitchell, P. (2006). Virtual environments for social skills training: Comments from two adolescents with autistic spectrum disorder. *Computers and Education*, 47, 186–206. DOI: 10.1016/j.compedu.2004.10.003. 88

Parsons, S., Mitchell, P., and Leonard, A. (2004). The use and understanding of virtual environments by adolescents with autistic spectrum disorders. *Journal of Autism and Developmental Disorders*, 34, 449–466. DOI: 10.1023/B:JADD.0000037421.98517.8d. 85, 87, 88

Parsons, S., Mitchell, P., and Leonard, A. (2005). Do adolescents with autistic spectrum disorders adhere to social conventions in virtual environments? *Autism*, 9, 95–117. DOI: 10.1177/1362361305049032. 87, 88

Parsons, S., Newbutt, N., and Wallace, S. (2013). Using virtual reality technology to support the learning of children on the autism spectrum. In K. Boser, M. Goodwin, and S. Wayland (Eds.) *Technology Tools for Students with Autism: Innovations that Enhance Independence and Learning*. Brookes. 85, 103

Parsons, T. D., Bowerly, T., Buckwalter, J. G., and Rizzo, A. A. (2007). A controlled clinical comparison of attention performance in children with ADHD in a virtual reality classroom compared to standard neuropsychological methods. *Child Neuropsychology*, 13(4), 363–381. DOI: 10.1080/13825580600943473.

Parsons, S., Yuill, N., Good, J., and Brosnan, M. (2020). 'Whose agenda? Who knows best? Whose voice?' Co-creating a technology research roadmap with autism stakeholders. *Disability & Society*, 35(2), 201-234. DOI: 10.1080/09687599.2019.1624152. 12

Peca, A., Simut, R., Pintea, S., Costescu, C., and Vanderborght, B. (2014). How do typically developing children and children with autism perceive different social robots?. *Computers in Human Behavior*, 41, 268–277. DOI: 10.1016/j.chb.2014.09.035. 143

Pennington, R. C., Stenhoff, D. M., Gibson, J., and Ballou, K. (2012). Using simultaneous prompting to teach computer-based story writing to a student with autism. *Education and Treatment of Children*, 35(3), 389–406. DOI: 10.1353/etc.2012.0022. 36

Picard, R. W. (2009). Future affective technology for autism and emotion communication. *Philosophical Transactions of the Royal Society B: Biological Sciences*, 364(1535), 3575–3584. DOI: 10.1098/rstb.2009.0143. 128

Picard, R. W. (2010). Emotion research by the people, for the people. *Emotion Review*, 2, 250–254. DOI: 10.1177/1754073910364256.

Pierno, A. C., Mari, M., Lusher, D., and Castiello, U. (2008). Robotic movement elicits visuomotor priming in children with autism. *Neuropsychologia*, 46, 448–454. DOI: 10.1016/j.neuropsychologia.2007.08.020. 136

Pinder, R. (1996). Sick-but-fit or Fit-but-sick? Ambiguity and identity in the workplace. In C. Barnes, and G. Mercer (Eds.), *Exploring the Divide* (pp. 135–156). Leeds: The Disability Press. 10

Pioggia, G., Ferro, M., Sica, M. L., Dalle Mura, G., Igliozzi, F., Muratori, S., and Casalini, S. (2006). Imitation and learning of the emotional behaviour: toward an android-based treatment for people with autism. *Proceedings of the Sixth International Workshop Epigenetics Robotics*, Sept. 20–22, Paris, pp. 119–25. Lund, Swed.: LUCS. 135

Pioggia, G., Igliozzi, R., Ferro, M., Ahluwalia, A., Muratori, F., and De Rossi, D. (2005). An android for enhancing social skills and emotion recognition in people with autism. *Neural Systems and Rehabilitation Engineering, IEEE Transactions on*, 13(4), 507–515. DOI: 10.1109/TNSRE.2005.856076. 127, 134

Pioggia, G., Igliozzi, R., Sica, M. L., Ferro, M., Muratori, F., Ahluwalia, A., and De Rossi, D. (2008). Exploring emotional and imitational android-based interactions in autistic spectrum disorders. *Journal of CyberTherapy and Rehabilitation*, 1, 49–62. 134

Piper, A. M., O'Brien, E., Morris, M. R., and Winograd, T. (2006). SIDES: a cooperative tabletop computer game for social skills development. In *Proceedings of the 2006 20th Anniversary Conference on Computer Supported Cooperative Work*, 1–10. DOI: 10.1145/1180875.1180877. 12, 69, 75, 76, 78, 79

Plienis, A. J. and Romanczyk, R. G. (1985). Analyses of performance, behavior, and predictors for severely disturbed children: A comparison of adult vs. computer instruction. *Analysis and Intervention in Developmental Disabilities*, 5(4), 345–356. DOI: 10.1016/0270-4684(85)90004-7. 38

Plotz, T., Hammerla, N. Y., Rozga, A., Reavis, A., Call, N., and Abowd, G. D. (2012). Automatic assessment of problem behavior in individuals with developmental disabilities. *Proceedings of the 2012 ACM Conference on Ubiquitous Computing*, 391–400. DOI: 10.1145/2370216.2370276. 115

Politis, Y., Sung, C., Goodman, L., and Leahy, M. (2019). Conversation skills training for people with autism through virtual reality: using responsible research and innovation approach. *Advances in Autism*. DOI: 10.1108/AIA-05-2018-0017. 88

Pop, C. A., Simut, R., Pintea, S., Saldien, J., Rusu, A., David, D., Vanderfaeillie, J., Lefeber, D., and Vanderborght, B. (2013). Can the social robot Probo help children with autism to identify situation-based emotions? a Series of single case experiments. *International Journal of Humanoid Robotics*, 10(03), 1350025. DOI: 10.1142/S0219843613500254. 138

Porayska-Pomsta, K., Frauenberger, C., Pain, H., Rajendran, G., Smith, T., Menzies, R., and Lemon, O. (2012). Developing technology for autism: an interdisciplinary approach. *Personal and Ubiquitous Computing*, 16(2), 117–127. DOI: 10.1007/s00779-011-0384-2. 12

Pöttgen, J., Dziobek, I., Reh, S., Heesen, C., and Gold, S. M. (2013). Impaired social cognition in multiple sclerosis. *Journal of Neurology, Neurosurgery and Psychiatry*, 84(5), 523–528. DOI: 10.1136/jnnp-2012-304157. 50

Prabhakar, K., Oh, S., Wang, P., Abowd, G. D., and Rehg, J. M. (2010). Temporal causality for the analysis of visual events. In *Computer Vision and Pattern Recognition* (CVPR), 2010 IEEE Conference on (pp. 1967–1974). IEEE. DOI: 10.1109/CVPR.2010.5539871.

Pradhan, A., Mehta, K., and Findlater, L. (2018). Accessibility came by accident: use of voice-controlled intelligent personal assistants by people with disabilities. In *Proceedings of the 2018 CHI Conference on Human Factors in Computing Systems* (p. 459). ACM. DOI: 10.1145/3173574.3174033. 130

Prince, E. B., Kim, E. S., Wall, C. A., Gisin, E., Goodwin, M. S., Simmons, E. S., Chawarska, K., and Shic, F. (2017). The relationship between autism symptoms and arousal level in toddlers with autism spectrum disorder, as measured by electrodermal activity. *Autism*, 21(4), 504–508. DOI: 10.1177/1362361316648816. 113

Putnam, C. and Chong, L. (2008). Software and technologies designed for people with autism: what do users want? In *Proceedings of the 10th International ACM SIGACCESS Conference on Computers and Accessibility* (pp. 3–10). ACM. DOI: 10.1145/1414471.1414475. 12

Pyles, D. A. M., Riordan, M. M., and Bailey, J. S. (1997). The stereotypy analysis: An instrument for examining environmental variables associated with differential rates of stereotypic behavior. *Research in Developmental Disabilities*, 18, 11–38. DOI: 10.1016/ S0891-4222(96)00034-0. 114

Quill, K. A. (1997). Instructional considerations for young children with autism: The rationale for visually cued instruction. *Journal of Autism and Developmental Disorders*. 27:6, 697–714. DOI: 10.1023/A:1025806900162. 31

Rajendran, G. (2013). Virtual environments and autism: a developmental psycho-pathological approach. *Journal of Computer Assisted Learning*. DOI: 10.1111/jcal.12006. 86, 103Rajendran, G., Mitchell, P., and Rickards, H. (2005). How do individuals with Asperger syndrome respond to nonliteral language and inappropriate requests in computer-mediated communication?. *Journal of Autism and Developmental Disorders*, 35(4), 429–443. DOI: 10.1007/s10803-005-5033-z. 37

Ramdoss, S., Lang, R., Mulloy, A., Franco, J., O'Reilly, M., Didden, R., and Lancioni, G. (2011). Use of computer-based interventions to teach communication skills to children with autism spectrum disorders: A systematic review. *Journal of Behavioral Education*, 20(1), 55–76. DOI: 10.1007/s10864-010-9112-7. 12, 15

Ranfelt, A. M., Wigram, T., and Øhrstrøm, P. (2009). Toward a handy interactive persuasive diary for teenagers with a diagnosis of autism. In *Proceedings of the 4th International Conference on Persuasive Technology* (p. 3). ACM. DOI: 10.1145/1541948.1541953. 62

Rangel, C. and Tentori, M., (2011). Self-configurable activities in activity-aware computing: The case of autism. In *Proceedings of 5th International Symposium on Ubiquitous Computing and Ambient Intelligence* (UCAmI), Springer-Verlag. Riviera, Maya, Mexico, December 5–9, 2011. 44, 62

Rani, N.M. and Ramli, S.H. and Legino, Rafeah and Azahari, M.H.H and Kamaruzaman, Muhamad. (2016). Comparative study on the engagement of students with autism towards learning through the use of mobile technology based visual schedule. Pp. 132–138. Available at https://tinyurl.com/y7dhf7cm.

Rathkey, J. H., Flax, S. W., Krug, D. A., and Arick, J. (1979). A microprocessor-based aid for training autistic children. *ISA Transactions*, 18(2), 79. 11

Reed, F. D., Hyman, S. R., and Hirst, J. M. (2011). Applications of technology to teach social skills to children with autism. *Research in Autism Spectrum Disorders*, 5(3), 1003–1010. DOI: 10.1016/j.rasd.2011.01.022. 12

Reed, P. (ed.) (1997). *Designing Environments for Successful Kids: A Resource Manual*. Wisconsin Assistive Technology Initiative (WATI). 69

Rehfeldt, R. A., Kinney, E. M., Root, S., and Stromer, R. (2004). Creating activity schedules using Microsoft® PowerPoint®. *Journal of Applied Behavior Analysis*, 37(1), 115–128. DOI: 10.1901/jaba.2004.37-115. 40

Rehg, J. M., Rozga, A., Abowd, G. D., and Goodwin, M. S. (2014). Behavioral imaging and autism. *IEEE Pervasive Computing*, 13(2), 84–87. DOI: 10.1109/MPRV.2014.23. 111

Rehg, J., Abowd, G., Rozga, A., Romero, M., Clements, M. R. M., Sclaroff, S. Essa, I., Ousley, O., Li, Y., Kim, C., Rao, H., Kim, J., Lo Presti, L., Zhang, J., Lantsman, D., Bidwell, J., and Ye, Z. (2013). Decoding children's social behavior. In *Proceedings of the 2013 IEEE Conference on Computer Vision and Pattern Recognition* (CVPR). IEEE. DOI: 10.1109/CVPR.2013.438.

Ricks, D. J. and Colton, M. B. (2010). Trends and considerations in robot-assisted autism therapy. *Proceedings - IEEE International Conference on Robotics and Automation*, pp. 4354–4359. DOI: 10.1109/ROBOT.2010.5509327. 142

Riedl, M., Arriaga, R., Boujarwah, F., Hong, H., Isbell, J., and Heflin, L. J. (2009). Graphical social scenarios: Toward intervention and authoring for adolescents with high functioning autism. *Virtual Healthcare Interaction: Papers from the AAAI Fall Symposium* (pp. 64–73). 45

Riek, L. D. (2012). Wizard of Oz studies in HRI: A systematic review and new reporting guidelines. *Journal of Human-Robot Interactions*, 1, 119–136. DOI: 10.5898/ JHRI.1.1.Riek. 133

Rimland, E. R. (1964). *Infantile Autism: The Syndrome and Its Implications for a Neural Theory of Behavior*. New York: Appleton-Century-Crofts. 2

Ringland, K. E. (2018). *Playful Places in Online Playgrounds: An Ethnography of a Minecraft Virtual World for Children with Autism*. UC Irvine. ProQuest ID: Ringland_uci_0030D_15167. 89

Ringland, K. E., Wolf, C. T., Boyd, L. E., Brown, J., Palermo, A., Lakes, K., and Hayes, G. R. (2019). DanceCraft: A whole-body interactive system for children with autism. *ASSETS 2019*. ACM. Pittsburgh, PA. DOI: 10.1145/3308561.3354604. 92

Ringland, K. E., Wolf, C. T., Faucett, H., Dombrowski, L., and Hayes, G. R. (2016). Will I always be not social?: Re-Conceptualizing sociality in the context of a Minecraft Community for autism. In *Proceedings of the 2016 CHI Conference on Human Factors in Computing Systems* (pp. 1256–1269). ACM. DOI: 10.1145/2858036.2858038. 89

Ringland, K. E., Zalapa, R., Neal, M., Escobedo, L., Tentori, M., and Hayes, G. R. (2014). SensoryPaint: a multimodal sensory intervention for children with neurodevelopmental disorders. In *Proceedings of the 2014 ACM International Joint Conference on Pervasive and Ubiquitous Computing* (pp. 873–884). ACM. DOI: 10.1145/2632048.2632065. 98

Ríos-Rincón, A. M., Adams, K., Magill-Evans, J., and Cook, A. (2016). Playfulness in children with limited motor abilities when using a robot. *Physical and Occupational Therapy in Pediatrics*, 36(3), 232–246. DOI: 10.3109/01942638.2015.1076559. 140

Rizzo, A. A., Bowerly, T., Buckwalter, J.G., Klimchuk, D., Mitura, R., and Parsons, T.D. (2006). A virtual reality scenario for all seasons: The virtual classroom. *CNS Spectrums*, 11(1), 35–44. DOI: 10.1017/S1092852900024196. 85

Rizzo, A. A., Buckwalter, J.-G., Bowerly, T., Van Der Zaag, C., Humphrey, L., Neumann, U., Chua, C., Kyriakakis, C., Van Rooyen, A., and Sisemore, D. (2000). The virtual classroom: A virtual reality environment for the assessment and rehabilitation of attention deficits. *CyberPsychology and Behavior*, 3(3), 483–499. DOI: 10.1089/10949310050078940.

Robins ,B., Dautenhahn, K., and Dickerson, P. (2009). From isolation to communication: a case study evaluation of robot assisted play for children with autism with a minimally expressive humanoid robot. *Proceedings of the Second International Conference on Advances in Computer-Human Interactions*, Feb. 1–7, Cancun, Mex., pp. 205–11. Piscataway, NJ: IEEE. DOI: 10.1109/ACHI.2009.32. 135, 136, 137

Robins, B. and Dautenhahn, K. (2014). Tactile interactions with a humanoid robot: Novel play scenario implementations with children with autism. *International Journal of Social Robotics*, 6(3), 397–415. DOI: 10.1007/s12369-014-0228-0. 137

Robins, B., Dautenhahn, K., and Dubowski, J. (2006). Does appearance matter in the interaction of children with autism with a humanoid robot? *Interaction Studies*, 7(3), 509–542. DOI: 10.1075/is.7.3.16rob. 134

Robins, B., Dautenhahn, K., Boekhorst, R. Te, and Billard, A. (2005). Robotic assistants in therapy and education of children with autism: can a small humanoid robot help encourage social interaction skills? *Universal Access in the Information Society*, 4(2), 105–120. DOI: 10.1007/s10209-005-0116-3. 136

Robins, B., Dickerson, P., Stribling, P., and Dautenhahn, K. (2004). Robot-mediated joint attention in children with autism: a case study in robot-human interaction. *Interaction Studies* 5(2):161–98. DOI: 10.1075/is.5.2.02rob. 135

Rogers, S. (1999). An examination of the imitation deficits in autism. In J. Nadel and G. Butterworth (Eds.), *Imitation in Infancy: Cambridge Studies in Cognitive Perceptual Development* (pp. 254–283). New York: Cambridge University Press. 6, 136

Rogers, S. J. (1998). Empirically supported comprehensive treatments for young children with autism. *Journal of Clinical Child Psychology*, 27, 167–178. DOI: 10.1207/ s15374424jccp2702_4. 9

Rogers, S. J. (2000). Interventions that facilitate socialization in children with autism. *Journal of Autism and Developmental Disorders*: Special Issue: Treatments for people with autism and other pervasive developmental disorders: Research perspectives, 30, 399–409. DOI: 10.1023/A:1005543321840. 9

Rogers, S. J. and Bennetto, L. (2000). Intersubjectivity in autism: The roles of imitation and executive function. In A.M. Wetherby and B.M. Prizant (Eds.), *Autism Spectrum Disorders: A Transactional Developmental Perspective* (pp. 79–107). Baltimore, MD: Paul H. Brookes Publishing. 136

Rojahn, J., Matson, J., Lott, D., Esbensen, A., and Smalls, Y. (2001). The behavior problems inventory: An instrument for the assessment of self-injury, stereotyped behavior, and aggression/destruction in individuals with developmental disabilities. *Journal of Autism and Developmental Disorders*, 31(6):577–588. DOI: 10.1023/A:1013299028321. 114

Rose, F. D., Brooks, B. M., Rizzo, A. A., Liebert, M. A., and Rose, C. (2005). Virtual reality in brain damage rehabilitation: Review. *CyberPsychology and Behavior*, 8(3), 263–272. DOI: 10.1089/cpb.2005.8.241.

Rougnant, A., Gelsomini, M., and Garzotto, F. (2017). WAYZ: a co-designed touch-ready game for children with special needs to assess and improve visual perception skills. In *2017 IEEE 25th International Requirements Engineering Conference Workshops* (REW) (pp. 160–163). IEEE. DOI: 10.1109/REW.2017.40.

Roy, D., Patel, R., DeCamp, P., Kubat, R., Fleischman, M., Roy, B., Mavridis, N., Tellex, S., Salata, A., Guinness, J., Levit, M., and Gorniak, P. (2006). The human speechome project. In *Proceedings of the 28th Annual Cognitive Science Conference*, (pp. 2059–2064), Mahwah, NJ: Lawrence Earlbaum. http://csjarchive.cogsci.rpi.edu/Proceedings/2006/ docs/p2059. pdf. 110

Russell, G., Mandy, W., Elliott, D., White, R., Pittwood, T., and Ford, T. (2019). Selection bias on intellectual ability in autism research: a cross-sectional review and meta-analysis. *Molecular Autism*, 10(1), 9. DOI: 10.1186/s13229-019-0260-x. 117

Rutter, M. (1970). Autistic children: Infancy to adulthood. *Semin Psychiatry*, 2, 435–450. 2

Rutter, M. (1974). The development of infantile autism. *Psychological Medicine*, 4(2), 147–163. DOI: 10.1017/S0033291700041982. 6

Rutter, M. (2005). Incidence of autism spectrum disorders: Changes over time and their meaning. *Acta Paediatrica*, 94(1), 2–15. DOI: 10.1080/08035250410023124. 4

Rutter, M. and Schopler, E. (1987). Autism and pervasive developmental disorders: Concepts and diagnostic issues. *Journal of Autism and Developmental Disorders*, 17(2), 159–186. DOI: 10.1007/BF01495054. 11

Sahin, N., Keshav, N., Salisbury, J., and Vahabzadeh, A. (2018). Safety and lack of negative effects of wearable augmented-reality social communication aid for children and adults with autism. *Journal of Clinical Medicine*, 7(8), 188. DOI: 10.3390/jcm7080188. 94, 95

Salter, T., Dautenhahn, K., and Te Boekhorst, R. (2006). Learning about natural human-robot interaction styles. *Robotics and Autonomous Systems*, 54(2), 127–134. DOI: /10.1016/j.robot.2005.09.022. 138

Sansosti, F. J. and Powell-Smith, K. A. (2008). Using computer-presented social stories and video models to increase the social communication skills of children with high-functioning autism spectrum disorders. *Journal of Positive Behavior Interventions*, 10(3), 162–178. DOI: 10.1177/1098300708316259. 43

Sarabadani, S., Schudlo, L. C., Samadani, A. A., and Kushki, A. (2018). Physiological detection of affective states in children with autism spectrum disorder. *IEEE Transactions on Affective Computing*. DOI: 10.1109/TAFFC.2018.2820049. 113

Savvides, P., Tolentino, L., Johnson-Glenberg, M. C., and Birchfield, D. (2010). A mixed-reality game to support communication for students with autism. Poster presented at SITE: Society for Information Technology and Teacher Education, San Diego, CA. 93

Scassellati, B. (2007). How social robots will help us diagnose, treat, and understand autism. *Robotics Research*, 28, 552–563. DOI: 10.1007/978-3-540-48113-3_47. 134, 135

Scassellati, B., Boccanfuso, L., Huang, C. M., Mademtzi, M., Qin, M., Salomons, N., Ventolo, P., and Shic, F. (2018). Improving social skills in children with ASD using a long-term, in-home social robot. *Science Robotics*, 3(21). DOI: 10.1126/scirobotics.aat7544. 135

Scassellati, B.(2005). Quantitative metrics of social response for autism diagnosis. *Proceedings of the 14th IEEE International Workshop on Robot Human Interactions and Communication (RO-MAN 2005)*, Aug. 13–15, Nashville, TN, pp. 585–90. Piscataway, NJ: IEEE. DOI: 10.1109/ROMAN.2005.1513843. 134

Scassellatti, B., Admoni, H., and Mataric, M. J. (2012). Robots for use in autism research. *Annual Review of Biomedical Engineering*, 14, 275–294. DOI: 10.1146/annurev-bio- eng-071811-150036. 134, 136, 142

Schepis, M. M., Reid, D. H., Behrmann, M. M., and Sutton, K. A. (1998). Increasing communicative interactions of young children with autism using a voice output communication aid and naturalistic teaching. *Journal of Applied Behavior Analysis*, 31(4), 561–578. DOI: 10.1901/jaba.1998.31-561.

Schlosser, R. W. and Blischak, D. M. (2001). Is there a role for speech output in interventions for persons with autism? A review. *Focus on Autism and Other Developmental Disabilities*, 16(3), 170–178. DOI: 10.1177/108835760101600305. 12

Schmuckler, M. A. (2001). What is ecological validity? A dimensional analysis. *Infancy*, 2, 419–436. DOI: 10.1207/S15327078IN0204_02. 108

Schreibman, L. and Ingersoll, B. (2005). Behavioral interventions to promote learning in individuals with autism. In F. Volkmar, A., Klin, R. Paul, and D. Cohen (Eds.), *Handbook of Autism and Pervasive Developmental Disorders*, Volume 2, Assessment, Interventions, and Policy (pp. 882–896). New York: Wiley. DOI: 10.1002/9780470939352.ch8. 9

Schreibman, L., Whalen, C., and Stahmer, A. C. (2000). The use of video priming to reduce disruptive transition behavior in children with autism. *Journal of Positive Behavior Interventions*, 2, 3–11. DOI: 10.1177/109830070000200102. 41

Schultheis, M. and Rizzo, A. (2001). The application of virtual reality technology in rehabilitation. *Rehabilitation Psychology*, 46(3), 296–311. DOI: 10.1037/0090-5550.46.3.296.

Self, T., Scudder, R. R., Weheba, G., and Crumrine, D. (2007). A virtual approach to teaching safety skills to children with autism spectrum disorder. *Topics in Language Disorders*, 27, 242–253. DOI: 10.1097/01.TLD.0000285358.33545.79. 90

Seltzer, M. M., Shattuck, P., Abbeduto, L., and Greenberg, J. S. (2004). Trajectory of development in adolescents and adults with autism. *Mental Retardation and Developmental Disabilities Research Reviews*, 10, 234–247. DOI: 10.1002/mrdd.20038. 6, 7

Shams, L. and Seitz, A. R. (2008). Benefits of multisensory learning. *Trends in Cognitive Sciences*, 12(11), 411–417. DOI: 10.1016/j.tics.2008.07.006. 95

Shamsuddin, S., Yussof, H., Ismail, L., Hanapiah, F. A., Mohamed, S., Piah, H. A., and Zahari, N. I. (2012). Initial response of autistic children in human-robot interaction therapy with humanoid robot NAO. In *Signal Processing and its Applications* (CSPA), *2012 IEEE 8th International Colloquium on* (pp. 188–193). IEEE. DOI: 10.1109/CSPA.2012.6194716. 140

Shane, H. C., Laubscher, E. H., Schlosser, R. W., Flynn, S., Sorce, J. F., and Abramson, J. (2012). Applying technology to visually support language and communication in individuals with autism spectrum disorders. *Journal of Autism and Developmental Disorders*, 42(6), 1228–1235. DOI: 10.1007/s10803-011-1304-z.

Shanhong Liu. (2019). "Tablet ownership among U.S. adults 2010-2019." https://www.statista.com/statistics/756045/tablet-owners-among-us-adults/. 66

Sherer, M., Pierce, K. L., Paredes, S., Kisacky, K. L., Ingersoll, B., and Schreibman, L. (2001). Enhancing conversation skills in children with zutism via video technology which is better, "self " or "other" as a model? *Behavior Modification*, 25(1), 140–158. DOI: 10.1177/0145445501251008. 43

Shibata, T., Kawaguchi, Y., and Wada, K. (2012). Investigation on people living with seal robot at home. *International Journal of Social Robotics*, 4(1), 53–63. DOI: 10.1007/s12369-011-0111-1.

Shinohara, K. and Wobbrock, J. O. (2011). In the shadow of misperception: assistive technology use and social interactions. In *Proceedings of the SIGCHI Conference on Human Factors in Computing Systems* (pp. 705–714). ACM. DOI: 10.1145/1978942.1979044. 120

Shipley-Benamou, R., Lutzker, J., and Taubman, M. (2002). Teaching daily living skills to children with autism through instructional video modeling. *Journal of Positive Behavior Interventions*, 4(3), 165–175. DOI: 10.1177/10983007020040030501. 40, 41

Shore, R. (1997). Rethinking the Brain: New Insights into Early Development. Families and work Institute. New York, NY.

Shukla-Mehta, S., Miller, T., and Callahan, K. J. (2009). Evaluating the effectiveness of video instruction on social and communication skills training for children with autism spectrum disorders: A review of the literature. *Focus on Autism and Other Developmental Disabilities*, 25(1), 23–36. DOI: 10.1177/1088357609352901. 43

Shuren, J., Patel, B., and Gottlieb, S. (2018). FDA regulation of mobile medical apps. JAMA 320(4), 337–338. DOI: 10.1001/jama.2018.8832. 146

Shyman, E. (2016). The reinforcement of ableism: Normality, the medical model of disability, and humanism in applied behavior analysis and ASD. *Intellectual and Developmental Disabilities*, 54(5), 366–376. DOI: 10.1352/1934-9556-54.5.366. 32

Sigafoos, J. and Drasgow, E. (2001). Conditional use of aided and unaided AAC: A review and clinical case demonstration. *Focus on Autism and Other Developmental Disabilities*, 16(3), 152–161. DOI: 10.1177/108835760101600303.

Sigafoos, J., Green, V. A., Payne, D., Son, S. H., O'Reilly, M., and Lancioni, G. E. (2009). A comparison of picture exchange and speech-generating devices: Acquisition, preference, and effects on social interaction. *Augmentative and Alternative Communication*, 25(2), 99-109. DOI: 10.1080/07434610902739959. 57

Sigman, M. and Ruskin, E. (1999). Continuity and change in the social competence of children with autism, Down Syndrome, and developmental delays. Monographs of the Society for Research in Child Development, 64, 1–114. DOI: 10.1111/1540-5834.00002. 5

Silva, G. F. M., Raposo, A., and Suplino, M. (2014). Par: A collaborative game for multitouch tabletop to support social interaction of users with autism. *Procedia Computer Science*, 27, 84–93. DOI: 10.1016/j.procs.2014.02.011. 77

Silver, M. and Oakes, P. (2001). Evaluation of a new computer intervention to teach people with autism or Asperger syndrome to recognize and predict emotions in others. *Autism*, 5(3), 299–316. DOI: 10.1177/1362361301005003007. 11, 45

Simm, W., Ferrario, M. A., Gradinar, A., and Whittle, J. (2014). Prototyping "clasp": implications for designing digital technology for and with adults with autism. In *Proceedings of the 2014 Conference on Designing Interactive Systems* (pp. 345–354). ACM. DOI: 10.1145/2598510.2600880. 9

Simut, R. E., Vanderfaeillie, J., Peca, A., Van de Perre, G., and Vanderborght, B. (2016). Children with autism spectrum disorders make a fruit salad with Probo, the social robot: an interaction study. *Journal of Autism and Developmental Disorders*, 46(1), 113–126. DOI: 10.1007/s10803-015-2556-9. 138

Simut, R., Pop, C., Vanderborght, B., Saldien, J., Rusu, A., Pintea, S., and David, D. (2011). The huggable social robot Probo for social story telling for robot assisted therapy with ASD children. *Proceedings of the 3rd International Conference on Social Robotics* (ICSR 2011), 97–100. Retrieved from http://probo.vub.ac.be/publications/The Huggable Social Robot Probo for Social Story Telling for Robot Assisted Therapy with ASD Children.pdf. DOI: 10.1075/is.13.3.02van.

Sitdhisanguan, K., Chotikakamthorn, N., Dechaboon, A., and Out, P. (2012). Using tangible user interfaces in computer-based training systems for low-functioning autistic children. *Personal and Ubiquitous Computing*, 16(2), 143–155. DOI: 10.1007/s00779-011-0382-4. 123

Slevin, E. (1999). Multisensory environments: Are they therapeutic? A single-subject evaluation of the clinical effectiveness of a multisensory environment. *Journal of Clinical Nursing*, 8(1), 48–56. DOI: 10.1046/j.1365-2702.1999.00211.x.

Smith, C. J., Rozga, A., Matthews, N., Oberleitner, R., Nazneen, N., and Abowd, G. (2017). Investigating the accuracy of a novel telehealth diagnostic approach for autism spectrum disorder. *Psychological Assessment*, 29(3), 245. DOI: 10.1037/pas0000317. 56

Smith, T., Groen, A. D., and Wynn, J. W. (2000). Randomized trial of intensive early intervention for children with pervasive developmental disorder. *American Journal on Mental Retardation*, 105, 269–285. DOI: 10.1352/0895-8017(2000)105<0269:RTOIEI>2.0.CO;2. 9

Sobel, K. (2018). *Designing Technology for Inclusive Play.* University of Washington Dissertation. 74

Sobel, K., O'Leary, K., and Kientz, J. A. (2015). Maximizing children's opportunities with inclusive play: considerations for interactive technology design. In *Proceedings of the 14th International Conference on Interaction Design and Children* June 2015, (pp. 39-48). DOI: 10.1145/2771839.2771844. 10

Sobel, K., Rector, K., Evans, S., and Kientz, J. A. (2016). Incloodle: evaluating an interactive application for young children with mixed abilities. In *Proceedings of the 2016*

CHI Conference on Human Factors in Computing Systems (pp. 165–176). ACM. DOI: 10.1145/2858036.2858114. 74, 78, 79

Solomon, O. (2012). The uses of technology for and with children with autism spectrum disorders. In L'Abate, L. and Kaiser, D. A. *Handbook of Technology in Psychology, Psychiatry and Neurology: Theory, Research, and Practice.* Nova Science Publishers. 12

Son, S. H. (2006). Comparing two modes of AAC intervention for children with autism (Doctoral dissertation, University of Texas at Austin, 2005). Dissertation Abstracts International: Section A: Humanities and Social Sciences, 66, 2544. 57

Spiel, K., Frauenberger, C., Keyes, O., and Fitzpatrick, G. (2019). Agency of autistic children in technology research—A critical literature review. *ACM Transactions on Computer-Human Interaction* (TOCHI), 26(6), 1–40. DOI: 10.1145/3344919. 12

Spiel, K., Malinverni, L., Good, J., and Frauenberger, C. (2017). Participatory evaluation with autistic children. In *Proceedings of the 2017 CHI Conference on Human Factors in Computing Systems* (pp. 5755–5766). ACM. DOI: 10.1145/3025453.3025851. 10

Sprague, R. L. and Newell, K. M. (Eds.). (1996). *Stereotyped Movements: Brain and Behavior Relationships.* Washington, DC: American Psychological Association. DOI: 10.1037/10202-000. 114

Srinivasan, S. M., Lynch, K. A., Bubela, D. J., Gifford, T. D., and Bhat, A. N. (2013). Effect of interactions between a child and a robot on the imitation and praxis performance of typically devloping children and a child with autism: A preliminary study 1, 2. *Perceptual And Motor Skills* (June 2015), 130729083743001. DOI: 10.2466/15.10.Pms.116.3.

Stadele, N. D. and Malaney, L. a. (2001). The effects of a multisensory environment on negative behavior and functional performance on individuals with autism. *Journal of Undergraduate Research*, IV, 211–218. Retrieved from http://murphylibrary.uwlax.edu/digital/jur/2001/stadele-malaney.pdf.

Stager, M., Lukowicz, P., Perera, N., von Buren, T., Troster, G., and Starner, T. (2003). Soundbutton: Design of a low power wearable audio classification system. In *Proceedings of the 7th IEEE International Symposium on Wearable Computers* (ISWC'03) (pp. 12–17). Los Alamitos, CA: IEEE Computer Society Press. DOI: 10.1109/ISWC.2003.1241387. 107

Standen, P. J. and Brown, D. J. (2006). Virtual reality and its role in removing the barriers that turn cognitive impairments into intellectual disability. *Virtual Reality*, 10(3–4), 241–252. DOI: 10.1007/s10055-006-0042-6.

Stanton, C. M., Kahn, P. H. Jr., Severson, R. L., Ruckert, J. H., and Gill, B. T. (2008). Robotic animals might aid in the social development of children with autism. *Proceedings of the 3rd*

ACM/ IEEE International Conference on Human-Robot Interactions (HRI '08), March 12–15, Amsterdam, pp. 271–78. New York: ACM. DOI: 10.1145/1349822.1349858. 134

Starner, T. E. (2001). The challenges of wearable computing: Part 1. In *IEEE Wearable Computing* (pp. 44–52). Los Alamitos, CA: IEEE Computer Society Press. DOI: 10.1109/40.946681. 107

Starner, T. E. (2002). Wearable computers: No longer science fiction. In *IEEE Pervasive Computing*, 1, 86–88. DOI: 10.1109/MPRV.2002.993148. 107

Stedman, A., Taylor, B., Erard, M., Peura, C., and Siegel, M. (2019). Are children severely affected by autism spectrum disorder underrepresented in treatment studies? An analysis of the literature. *Journal of Autism and Developmental Disorders*, 49(4), 1378–1390. DOI: 10.1007/s10803-018-3844-y. 108

Stendal, K. and Balandin, S. (2015). Virtual worlds for people with autism spectrum disorder: a case study in Second Life. *Disability and Rehabilitation*, 37(17), 1591-1598. DOI: 10.3109/09638288.2015.1052577. 38

Stephenson, J. and Carter, M. (2011). The use of multisensory environments in schools for students with severe disabilities: Perceptions from teachers. *Journal of Developmental and Physical Disabilities*, 23(4), 339–357. DOI: 10.1007/s10882-011-9232-6. 97

Stiehl, W. D., Lieberman, J., Breazeal, C., Basel, L., Lalla, L., and Wolf, M. (2005). Design of a therapeutic robotic companion for relational, affective touch. In *Robot and Human Interactive Communication*, 2005. *ROMAN 2005. IEEE International Workshop on* (pp. 408–415). IEEE. DOI: 10.1109/ROMAN.2005.1513813. 138

Stone, W. L., Ousley, O. Y., Yoder, P. J., Hogan, K. L., and Hepburn, S. L. (1997). Nonverbal communication in two- and three-year-old children with autism. *Journal of Autism and Developmental Disorders*, 27, 677–696. DOI: 10.1023/A:1025854816091. 5

Strasberger, S. K., and Ferreri, S. J. (2014). The effects of peer assisted communication application training on the communicative and social behaviors of children with autism. *Journal of Developmental and Physical Disabilities*, 26(5), 513-526. DOI: 10.1007/s10882-013-9358-9. 58

Strickland, D. (1996). A virtual reality application with autistic children. *Presence: Tel-eoperators and Virtual Environments*, 5, 319–329. DOI: 10.1162/pres.1996.5.3.319. 85

Strickland, D. (1997). Virtual reality for the treatment of autism. *Studies in Health Technology and Informatics*, 44, 81–86. DOI: 10.3233/978-1-60750-888-5-81.

Strickland, D. (1998). Virtual reality for the treatment of autism. In G. Riva (Ed.), *Virtual Reality in Neuro-Psycho-Physiology* (pp. 81–86). Amsterdam, Netherlands: Ios Press, 1998. 85, 87

Strickland, D. C. D. D. C., McAllister, D., Coles, C. D. C., and Osborne, S. (2007). An evolution of virtual reality training designs for children with autism and fetal alcohol spectrum disorders. *Topics in Language Disorders*, 27(3), 226–241. DOI: 10.1097/01. TLD.0000285357.95426.72. 90

Strickland, D., Marcus, L. M., Mesibov, G. B., and Hogan, K. (1996). Brief report: Two case studies using virtual reality as a learning tool for autistic children. *Journal of Autism and Developmental Disorders*, 26(6), 651–659. DOI: 10.1007/BF02172354. 85, 87

Stromer, R., Kimball, J. W., Kinney, E. M., and Taylor, B. A. (2006). Activity sched ules, computer technology, and teaching children with autism spectrum disor ders. *Focus on Autism and Other Developmental Disabilities*, 21(1), 14–24. DOI: 10.1177/10883576060210010301. 40, 44

Suh, H. (2019). Supporting Long-Term Health Monitoring by Reducing User Burden and Enhancing User Benefit. Ph.D. Dissertation. University of Washington.

Suh, H., Porter, J. R., Hiniker, A., and Kientz, J. A. (2014). @ BabySteps: design and evaluation of a system for using twitter for tracking children's developmental milestones. In *Proceedings of the 32nd annual ACM Conference on Human Factors in Computing Systems* (pp. 2279–2288). ACM. DOI: 10.1145/2556288.2557386. 35

Suh, H., Porter, J. R., Racadio, R., Sung, Y. C., and Kientz, J. A. (2016). Baby steps text: Feasibility study of an SMS-based tool for tracking children's developmental progress. In *AMIA Annual Symposium Proceedings* (Vol. 2016, p. 1997). American Medical Informatics Association. 35, 63, 64

Sung, A. N., Bai, A., Bowen, J., Xu, B., Bartlett, L. M., Sanchez, J. C., Chin, M. D., Poirier, L. J.., Blinkhorn, M. R., Campbell, A. C., and Tanaka, J. W. (2015). From the small screen to the big world: mobile apps for teaching real-world face recognition to children with autism. *Advanced Health Care Technologies*, 1, 37–45. DOI: 10.2147/AHCT.S64483. 59, 60

Sveistrup, H. (2004). Motor rehabilitation using virtual reality. *Journal of Neuroengineering and Rehabilitation*, 1, 10. DOI: 10.1186/1743-0003-1-10.

Sverd, J., Montero, G., and Gurevich, N. (1993). Cases for an association between Tourette syndrome, autistic disorder, and schizophrenia-like disorder. *Journal of Autism and Developmental Disorders*, 23, 407–413. DOI: 10.1007/BF01046229. 3

Swanson, M. R., Shen, M. D., Wolff, J. J., Boyd, B., Clements, M., Rehg, J., Elison, J. T., Paterson, S., Parish-Morris, J., Chappell, J. C., Hazlett, H. C., Emerson, R. W., Botteron, K., Pandey,

J., Schultz, R. T., Dager. S. R., Zwaigenbaum, L., Estes, A. M., Piven, J., and IBIS Network. (2018). Naturalistic language recordings reveal "hypervocal" infants at high familial risk for autism. *Child Development*, 89(2), e60–e73. DOI: 10.1111/cdev.12777. 112

Syahputra, M. F., Arisandi, D., Lumbanbatu, A. F., Kemit, L. F., Nababan, E. B., and Sheta, O. (2018). Augmented reality social story for autism spectrum disorder. *Journal of Physics: Conference Series*, 978(1), p. 012040). IOP Publishing. DOI: 10.1088/1742-6596/978/1/012040. 92, 93

Tager-Flusberg, H., Edelson, L., and Luyster, R. (2011). Language and communication in autism spectrum disorders. In D. G. Amaral, G. Dawson, and D. H. Geschwind (Eds.) *Autism Spectrum Disorders* (pp. 172–185). Oxford University Press. DOI: 10.1093/med/9780195371826.001.0001. 112

Tager-Flusberg, H., Paul, R. and Lord, C. (2005). Language and communication in autism. In F. Volkmar, A., Klin, and R. Paul (Eds.) *Handbook of Autism and Pervasive Developmental Disorders*. (pp. 335–364). New York: Wiley and Sons. DOI: 10.1002/9780470939345. ch12. 5

Tam, V., Gelsomini, M., and Garzotto, F. (2017). Polipo: A tangible toy for children with neurodevelopmental disorders. In *Proceedings of the Eleventh International Conference on Tangible, Embedded, and Embodied Interaction* (pp. 11–20). ACM. DOI: 10.1145/3024969.3025006. 124, 125 , 128, 129

Tanaka, J. W., Lincoln, S., and Hegg, L. (2003). A framework for the study and treatment of face processing deficits in autism. In H. Leder and G. Swartzer (Eds.), *The Development of Face Processing* (pp. 101–119). Berlin: Hogrefe. 32, 33

Tapus, A., Mataric, M. J., and Scassellati, B. (2007). The grand challenges in socially assistive robotics. *IEEE Robotic Automation Magazine Special Issue Grand Challenges in Robotics* 14(1):35–42. DOI: 10.1109/MRA.2007.339605. 133, 134

Tapus, A., Peca, A., Aly, A., Pop, C., Jisa, L., Pintea, S., Rusu, A. S., and David, D. O. (2012). Children with autism social engagement in interaction with Nao, an imitative robot: A series of single case experiments. *Interaction Studies*, 13(3), 315–347. DOI: 10.1075/is.13.3.01tap.

Tardif, C., Laine, F., Rodriguez, M., and Gepner, B. (2007). Slowing down presentation of facial movements and vocal sounds enhances facial expression recognition and induces facial–vocal imitation in children with autism. *Journal of Autism and Developmental Disorders*, 37(8), 1469–1484. DOI: 10.1007/s10803-006-0223-x. 44

Tartaro, A. and Cassell, J. (2008). Playing with virtual peers: Bootstrapping contingent discourse in children with autism. *Proceedings of International Conference of the Learning Sciences (ICLS)*, June 24–28, Utrecht, Netherlands. DOI: 10.5555/1599871.1599919. 88, 101

Taylor, B. A. and Levin, L. (1998). Teaching a student with autism to make verbal initiations: Effects of a tactile prompt. *Journal of Applied Behavior Analysis*, 31(4), 651–654. DOI: 10.1901/jaba.1998.31-651. 123

Taylor, B. A., Levin, L., and Jasper, S. (1999). Increasing play-related statements in children with autism toward their siblings: Effects of video modeling. *Journal of Developmental and Physical Disabilities*, 11(3), 253–264. DOI: 10.1023/A:1021800716392. 43

Tentori, M. and Hayes, G. R. (2010). Designing for interaction immediacy to enhance social skills of children with autism. In *Proceedings of the 12th ACM International Conference on Ubiquitous Computing* (pp. 51–60). ACM. DOI: 10.1145/1864349.1864359. 62

Tentori, M., Escobedo, L., and Balderas, G. (2015). A smart environment for children with autism. *IEEE Pervasive Computing*, 14(2), 42–50. DOI: 10.1109/MPRV.2015.22.

Thompson, C. J. (2011). Multi-sensory intervention observational research. *International Journal of Special Education*, 26(1), 202–214. DOI: 10.1016/j.biocon.2003.09.001.96

Tomasello, M. (1995). Joint attention as social cognition. In C. Moore and P. Dunham (Eds.), *Joint Attention: Its Origins and Role in Development* (pp. 103–130). Hillsdale, NJ: Erlbaum. 135

Tomasello, M., Kruger, A., and Ratner, H. (1993). Cultural learning. *Behavioral and Brain Sciences*, 16, 450–488. DOI: 10.1017/S0140525X0003123X. 6

Tordjman S., Anderson, G. M., Botbol, M., Brailly-Tabard, S., Perez-Diaz, F., Graignic, R., Carlier, M., Schmit, G., Rolland, A.-C., Bonnot, O., Trabado, S., Roubertoux, P., and Bronsard, G. (2009) Pain reactivity and plasma b-endorphin in children and adolescents with autistic disorder. *PLoS ONE* 4(8): e5289. DOI: 10.1371/journal.pone.0005289. 113

Torres, N. A., Clark, N., Popa, D., and Ranatunga, I. (2012). Implementation of interactive arm playback behaviors of social robot zeno For autism spectrum disorder therapy. *PETRA2012*, June 6–8, Crete, Greece. DOI: 10.1145/2413097.2413124. 136

Torta, E., Oberzaucher, J., Werner, F., Cuijpers, R. H., and Juola, J. F. (2013). Attitudes towards socially aassistive robots in intelligent homes: Results from laboratory studies and field trials. *Journal of Human-Robot Interaction*, 1(2), 76–99. DOI: 10.5898/JHRI.1.2.Torta.

Trepagnier, C. Y., Sebrechts, M. M., Finkelmeyer, A., Stewart, W., Woodford, J., and Coleman, M. (2006). Simulating social interaction to address deficits of autistic spectrum disorder in children. *Cyberpsychology and Behavior*, 9(2), 213–217. DOI: 10.1089/ cpb.2006.9.213. 127

Trepagnier, C. G. (1999). Virtual environments for the investigation and rehabilitation of cognitive and perceptual impairments. *NeuroRehabilitation*, 12, 63–72. DOI: 10.3233/NRE-1999-12107. 85

Truong, K. N., Abowd, G. D., and Brotherton, J. A. (2001). Who, what, when, where, how: design issues of capture and access applications. In *Proceedings of Ubicomp 2001*. 209–224. DOI: 10.1007/3-540-45427-6_17. 47

Tseng, R. Y. and Do, E. Y. L. (2010). Facial expression wonderland (FEW): a novel design prototype of information and computer technology (ICT) for children with autism spectrum disorder (ASD). In *Proceedings of the 1st ACM International Health Informatics Symposium* (pp. 464–468). ACM. DOI: 10.1145/1882992.1883064. 33

Turnacioglu, S., McCleery, J. P., Parish-Morris, J., Sazawal, V., and Solorzano, R. (2019). The state of virtual and augmented reality therapy for autism spectrum disorder (ASD). In *Virtual and Augmented Reality in Mental Health Treatment* (pp. 118–140). IGI Global. DOI: 10.4018/978-1-5225-7168-1.ch008. 87

Ulgado, R.R., Nguyen, K., Custodio, V.E., Waterhouse, A., Weiner, R. and Hayes, G.R. (2013) VidCoach: a mobile video modeling system for youth with special needs. In *Proceedings of the 12th International Conference on Interaction Design and Children* (IDC '13). 581–584. DOI: 10.1145/2485760.2485870. 43, 62

Uljarević, M., Hedley, D., Rose-Foley, K., Magiati, I., Cai, R. Y., Dissanayake, C., Richdale, A., and Trollor, J. (2019). Anxiety and depression from adolescence to old age in autism spectrum disorder. *Journal of Autism and Developmental Disorders*, 1–11. DOI: 10.1007/s10803-019-04084-z. 8

Van Laarhoven, T., Kraus, E., Karpman, K., Nizzi, R., and Valentino, J. (2010). A comparison of picture and video prompts to teach daily living skills to individuals with autism. *Focus on Autism and Other Developmental Disabilities*, 25(4), 195–208. DOI: 10.1177/1088357610380412. 42

van Santen, J., Prud'Homeaux, E. T., Black, L. M., and Mitchell, M. (2010). Computational prosodic markers for autism. *Autism*, 14(3), 215–236. DOI: 10.1177/1362361310363281. 112

van Santen, J., Sproat, R. W., and Presmanes Hill, A. (2013). Quantifying repetitive speech in autism spectrum disorders and language impairment. *Autism Research*, 6 (5) 372–383. DOI: 10.1002/aur.1301.112

van Schalkwyk, G. I., Marin, C. E., Ortiz, M., Rolison, M., Qayyum, Z., McPartland, J. C., Lebowitz, E. R., Volkmar, F. R., and Silverman, W. K. (2017). Social media use, friendship quality, and the moderating role of anxiety in adolescents with autism spectrum disorder. *Journal of Autism and Developmental Disorders*, 47(9), 2805–2813. DOI: 10.1007/s10803-017-3201-6. 36

van Steensel, F. J. A., Bogels, S. M., and Perrin, S. (2011). Anxiety disorders in children and adolescents with autistic spectrum disorders: A meta-analysis. *Clinical Child and Family Psychology Review* 14(3): 302–317. DOI: 10.1007/s10567-011-0097-0. 113

van Veen M, de Vries A, Cnossen F. (2009). Improving collaboration for children with PDD-NOS using a serious game with multi-touch interaction. In *Proceedings of CHI NL 2009*; Leiden, the Netherlands. 17–20. 72, 74, 75

Van Veen, M., De Vries, A., Cnossen, F., and Willems, R. (2009). Improving collaboration skills for children with PDD-NOS through a multi-touch based serious game. In *Proceedings of International Conference of Education and New Learning Technologies* (pp. 3559–70).

Van Den Heuvel, R. J., Lexis, M. A., Janssens, R. M., Marti, P., and De Witte, L. P. (2017). Robots supporting play for children with physical disabilities: exploring the potential of IROMEC. *Technology and Disability*, 29(3), 109-120. DOI: 10.3233/TAD-160166. 140

Vanderborght, B., Simut, R., Saldien, J., Pop, C., Rusu, A. S., Pintea, S., Lefeber, D., and David, D. O. (2012). Using the social robot probo as a social story telling agent for children with ASD. *Interaction Studies*, 13(3), 348–372. DOI: 10.1075/is.13.3.02van.

Vandevelde, C., Wyffels, F., Vanderborght, B., and Saldien, J. (2017). DIY design for social robots. *IEEE Robotics and Automation Magazine*, (March), 10. DOI: 10.1109/MRA.2016.2639059.

Vera, L., Campos, R., Herrera, G., and Romero, C. (2007). Computer graphics applications in the education process of people with learning difficulties. *Computers and Graphics*, 31, 649–658. DOI: 10.1016/j.cag.2007.03.003. 85

Vertegaal, R., Slagter, R., van der Veer, G., and Nijholt, A. (2001). Eye gaze patterns in conversations: There is more to conversational agents than meets the eyes. In *Proceedings of Human Factors in Computing Systems* (CHI 2001) (pp. 301–308). New York: ACM Press. DOI: 10.1145/365024.365119.

Villafuerte, L., Markova, M., and Jorda, S. (2012). Acquisition of social abilities through musical tangible user interface: Children with autism spectrum condition and the reactable. In *CHI'12 Extended Abstracts on Human Factors in Computing Systems* (pp. 745–760). ACM. DOI: 10.1145/2212776.2212847. 57

Vlaskamp, C., De Geeter, K. I., Huijsmans, L. M., and Smit, I. H. (2003). Passive activities: The effectiveness of multisensory environments on the level of activity of individuals with profound multiple disabilities. *Journal of Applied Research in Intellectual Disabilities*, 16(2), 135–143. DOI: 10.1046/j.1468-3148.2003.00156.x.

Volioti, C., Tsiatsos, T., Mavropoulou, S., and Karagiannidis, C. (2014). VLSS - Virtual learning and social stories for children with autism. *Proceedings of the IEEE 14th International Conference on Advanced Learning Technologies*, ICALT 2014, 606–610. DOI: 10.1109/ICALT.2014.177.

Volioti, C., Tsiatsos, T., Mavropoulou, S., and Karagiannidis, C. (2016). VLEs, social stories and children with autism: A prototype implementation and evaluation. *Education and Information Technologies*, 21(6), 1679–1697. DOI: 10.1007/s10639-015-9409-1.

Volkmar, F. R. and Nelson, D. S. (1990). Seizure disorders in autism. *Journal of the American Academy of Child and Adolescent Psychiatry*, 29, 127–129. DOI: 10.1097/00004583-199001000-00020. 3

Volkmar, F. R., Stier, D. M., and Cohen, D. J. (1985). Age of recognition of pervasive developmental disorder. *American Journal of Psychiatry*, 142:1450, 1452. DOI: 10.1176/ajp.142.12.1450. 109

Vosoughi, S., Goodwin, M. S., Washabaugh, B., and Roy, D. (2012). Speechome recorder for the study of child language development and disorders. *ACM International Conference on Multimodal Interaction*, 193–200. DOI: 10.1145/2388676.2388715. 110

Waddington, H., Sigafoos, J., Lancioni, G. E., O'Reilly, M. F., Van der Meer, L., Carnett, A., Stevens, M., Roche, L., Hodis, F., Green, V. A., Sutherland, D., Lang, R., and Marschik, P. B. (2014). Three children with autism spectrum disorder learn to perform a three-step communication sequence using an iPad®-based speech-generating device. *International Journal of Developmental Neuroscience*, 39, 59-67. DOI: 10.1016/j.ijdevneu.2014.05.001. 58

Wade, J., Zhang, L., Bian, D., Fan, J., Swanson, A., Weitlauf, A., Sarkar, M., Warren, Z., and Sarkar, N. (2016). A gaze-contingent adaptive virtual reality driving environment for intervention in individuals with autism spectrum disorders. *ACM Transactions on Interactive Intelligent Systems* (TiiS), 6(1), 3. DOI: 10.1145/2892636. 90

Wainer, A. L. and Ingersoll, B. R. (2011). The use of innovative computer technology for teaching social communication to individuals with autism spectrum disorders. *Research in Autism Spectrum Disorders*, 5(1), 96–107. DOI: 10.1016/j.rasd.2010.08.002. 12

Wainer, J., Dautenhahn, K., Robins, B., and Amirabdollahian, F. (2014). A pilot study with a novel setup for collaborative play of the humanoid robot KASPAR with children with autism.

International Journal of Social Robotics, 6(1), 45–65. DOI: 10.1007/s12369-013-0195-x. 137

Wainer, J., Ferrari, E., Dautenhahn, K., and Robins, B. (2010). The effectiveness of using a robotics class to foster collaboration among groups of children with autism in an exploratory study. *Personal and Ubiquitous Computing*, 14(5), 445–455. DOI: 10.1007/s00779-009-0266-z. 136

Wallace, S., Parsons, S., Westbury, A., White, K., White, K., and Bailey, A. (2010). Sense of presence and atypical social judgments in immersive virtual reality: Responses of adolescents with autistic spectrum disorders. *Autism*, 14(3), 199–213. DOI: 10.1177/1362361310363283. 87, 88, 101, 102

Wang, M. and Reid, D (2011). Virtual reality in pediatric neurorehabilitation: Attention deficit hyperactivity disorder, autism and cerebral palsy. *Neuroepidemiology*, 36, 2–18. DOI: 10.1159/000320847. 87, 103

Wang, P., Abowd, G. D., and Rehg, J. M. (2009). Quasi-periodic event analysis for social game retrieval. In *Computer Vision, 2009 IEEE 12th International Conference on* (pp. 112–119). IEEE. DOI: 10.1109/ICCV.2009.5459151.

Ward, D. M., Dill-Shackleford, K. E., and Mazurek, M. O. (2018). Social media use and happiness in adults with autism spectrum disorder. *Cyberpsychology, Behavior, and Social Networking*, 21(3), 205–209. DOI: 10.1089/cyber.2017.0331. 36

Warren, Z. E., Zheng, Z., Swanson, A. R., Bekele, E., Zhang, L., Crittendon, J. A., Weitlauf, A. F., and Sarkar, N. (2015). Can robotic interaction improve joint attention skills? *Journal of Autism and Developmental Disorders*, 45(11), 3726–3734. DOI: 10.1007/s10803-013-1918-4.

Warren, Z., Zheng, Z., Das, S., Young, E. M., Swanson, A., Weitlauf, A., and Sarkar, N. (2015). Brief report: Development of a robotic intervention platform for young children with ASD. *Journal of Autism and Developmental Disorders*, 45(12), 3870–3876. DOI: 10.1007/s10803-014-2334-0.

Weiner, N. (1948). *Cybernetics, or Communication and Control in the Animal and the Machine*. Cambridge: MIT Press.

Weiser, M. (1991). The computer for the 21st century. *Scientific American Special Issue on Communications, Computers, and Networks*, September, 1991. DOI: 10.1038/scientificamerican0991-94. 107

Weiss, P. L., Gal, E., Eden, S., Zancanaro, M., and Telch, F. (2011). Usability of a multitouch tabletop surface to enhance social competence training for children with autism spectrum disorder. In *Proceedings of Annual Conference of the Open University CHAIS Center*,71–78. DOI: 10.1109/ICVR.2011.5971867. 78

Welch, W. C., Lahiri, U., Warren, Z., and Sarkar, N. (2010). An approach to the design of socially acceptable robots for children with autism spectrum disorders. *International Journal of Social Robotics* 2:391–403. DOI: 10.1007/s12369-010-0063-x.

Werry, I., Dautenhahn, K., Ogden, B., and Harwin, W. (2001). Can social interaction skills be taught by a social agent? The role of a robotic mediator in autism therapy. In *Lecture Notes in Computer Science*,Vol. 2117: *Cognitive Technology: Instruments of Mind*, eds. M. Beynon, C. L. Nehaniv, and K. Dautenhahn, pp. 57–74. Berlin: Springer. DOI: 10.1007/3- 540-44617-6_6. 136

Westeyn, T. L., Abowd, G. D., Starner, T. E., Johnson, J. M., Presti, P. W., and Weaver, K. A. (2012). Monitoring children's developmental progress using augmented toys and activity recognition. *Personal and Ubiquitous Computing*,16(2), 169–191. DOI: 10.1007/ s00779-011-0386-0. 125

Westlund, J. M. K., Jeong, S., Park, H. W., Ronfard, S., Adhikari, A., Harris, P. L., DeSteno, D., and Breazeal, C. L. (2017). Flat vs. expressive storytelling: young children's learning and retention of a social robot's narrative. *Frontiers in Human Neuroscience*, 11, 295. DOI: 10.3389/fnhum.2017.00295. 143

Wetherby, A. M., Yonclas, D. G., and Bryan, A. A. (1989). Communicative profiles of pre- school children with handicaps: Implications for early identification. *Journal of Speech and Hearing Disorders*, 54, 148–158. DOI: 10.1044/jshd.5402.148. 5

Whalen, C., Liden, L., Ingersoll, B., Dallaire, E. and Liden, S. (2006). Positive behavioral changes associated with the use of computer-assisted instruction for young children. *Journal of Speech and Language Pathology and Applied Behavior Analysis* 1(1): 11–26. DOI: 10.1037/ h0100182. 32, 69

Whalen, C., Moss, D., Ilan, A. B., Vaupel, M., Fielding, P., Macdonald, K., Cernich, S., and Symon, J. (2010). Efficacy of TeachTown: Basics computer-assisted intervention for the intensive comprehensive autism program in Los Angeles unified school district. *Autism*, 14(3), 179–197. DOI: 10.1177/1362361310363282. 32, 49, 50

Whitcomb, S. A., Bass, J. D., and Luiselli, J. K. (2011). Effects of a computer-based early reading program (Headsprout®) on word list and text reading skills in a student with autism. *Journal of Developmental and Physical Disabilities*,23(6), 491–499. DOI: 10.1007/ s10882-011-9240-6. 36

White, D. R., Camacho-Guerrero, J. A., Truong, K. N., Abowd, G. D., Morrier, M. J., Vekaria, P. C., and Gromala, D. (2003). Mobile capture and access for assessing language and social development in children with sutism. In *Proceedings of Ubicomp '03*. Seattle, WA: Springer-Verlag, 137–140.

White, S. W., Oswald, D., Ollendick, T., and Scahill, L. (2009). Anxiety in children and adolescents with autism spectrum disorders. *Clinical Psychology Review*, 29, 216–229. DOI: 10.1016/j.cpr.2009.01.003. 113

Whyte, E. M., Smyth, J. M., and Scherf, K. S. (2015). Designing serious game interventions for individuals with autism. *Journal of Autism and Developmental Disorders*, 45(12), 3820-3831. DOI: 10.1007/s10803-014-2333-1. 66

Wilhem, F. and Grossman, P. (2010). Emotions beyond the laboratory: Theoretical fundaments, study design, and analytical strategies for advanced ambulatory assessment. *Biological Psychology*, 84 (3) 552–569. DOI: 10.1016/j.biopsycho.2010.01.017. 107

Willemsen-Swinkels, S. H. N., Buitelaar, J. K., Dekker, M., and van Engeland, H. (1998). Subtyping stereotypic behavior in children: The association between stereotypic behavior, mood, and heart rate. *Journal of Autism and Developmental Disorders*, 28, 547–557. DOI: 10.1023/A:1026008313284.

Williams, C. (2013). Evaluating therapeutic engagement and expressive communication in immersive multimedia environments. *Lecture Notes in Computer Science* (Including Subseries Lecture Notes in Artificial Intelligence and Lecture Notes in Bioinformatics), 8010 LNCS(PART 2), 514–523. DOI: 10.1007/978-3-642-39191-0_56.

Williams, C., Wright, B., Callaghan, G., and Coughlan, B. (2002). Do children with autism learn to read more readily by computer assisted instruction or traditional book methods? A pilot study. *Autism*, 6(1), 71–91. Wisconsin Assistive Technology Initiative, Oshkosh, WI. DOI: h10.1177/1362361302006001006. 39

Wills, H. P., Mason, R., Huffman, J. M., and Heitzman-Powell, L. (2019). Implementing self-monitoring to reduce inappropriate vocalizations of an adult with autism in the workplace. *Research in Autism Spectrum Disorders*, 58, 9–18. DOI: 10.1016/j.rasd.2018.11.007. 8, 9

Wing, L. and Attwood, A. (1987). Syndromes of autism and atypical development. In D. Cohen and A. Donnellan, (Eds.), *Handbook of Autism and Pervasive Disorders*, (pp. 3–19). New York: John Wiley and Sons. 6

Wing, L. and Gould, J. (1979). Severe impairments of social interaction and associated abnormalities in children: Epidemiology and classification. *Journal of Autism and Developmental Disorders*, 9, 11–29. DOI: 10.1007/BF01531288. 3, 6

Wing, L. and Potter, D. (2002). The epidemiology of autistic spectrum disorders: Is the prevalence rising? *Mental Retardation and Developmental Disabilities Research Reviews*, 8, 151–161. DOI: 10.1002/mrdd.10029. 4

Winoto, P., Tang, T. Y., and Guan, A. (2016). I will help you pass the puzzle piece to your partner if this is what you want me to: The design of collaborative puzzle games to train chinese children with autism spectrum disorder joint attention skills. In *Proceedings of the The 15th International Conference on Interaction Design and Children* (pp. 601–606). ACM. DOI: 10.1145/2930674.2936012. 74, 75

Wong, S. K. and Tam, S. F. (2001). Effectiveness of a multimedia programme and therapist-instructed training for children with autism. *International Journal of Rehabilitation Research*, 24(4), 269–278. DOI: 10.1097/00004356-200112000-00003. 44

Woodard, C. R., Goodwin, M. S., Zelazo, P. R., Aube, D., Scrimgeour, M., Ostholthoff, T., and Brickley, M. (2012). A comparison of autonomic, behavioral, and parent-report measures of sensory sensitivity in young children with autism. *Research in Autism Spectrum Disorders*, 6, 1234–1246. DOI: 10.1016/j.rasd.2012.03.012. 113

Wright, C., Diener, M. L., Dunn, L., and Wright, S. D. (2011). SketchUp™: a technology tool to facilitate intergenerational family relationships for children with autism spectrum disorders (ASD). *Family and Consumer Sciences Research Journal*, 40(2), 135–149. DOI: 10.1111/j.1552-3934.2011.02100.x.

Xin, J. F. and Rieth, H. (2001). Video-assisted vocabulary instruction for elementary school students with learning disabilities. *Information Technology in Childhood Education Annual*, 2001(1), 87–103. 43

Xu, D., Richards, J. A., Gilkerson, J., Yapanel, U., Gray, S., and Hansen, J. (2009). Automatic childhood autism detection by vocalization decomposition with phone-like units. *The 2nd Workshop on Child, Computer and Interaction*, Cambridge, MA, November 5. DOI: 10.1145/1640377.1640382. 112

Yang, Y. D., Allen, T., Abdullahi, S. M., Pelphrey, K. A., Volkmar, F. R., and Chapman, S. B. (2018). Neural mechanisms of behavioral change in young adults with high-functioning autism receiving virtual reality social cognition training: A pilot study. *Autism Research*, 11(5), 713–725. DOI: 10.1002/aur.1941. 88

Yu, C., Shane, H., Schlosser, R. W., O'Brien, A., Allen, A., Abramson, J., and Flynn, S. (2018). An exploratory study of speech-language pathologists using the Echo Show™ to deliver visual supports. *Advances in Neurodevelopmental Disorders*, 2(3), 286–292. DOI: 10.1007/s41252-018-0075-3. 126

Zancanaro, M., Giusti, L., Gal, E., and Weiss, P. T. (2011). Three around a table: the facilitator role in a co-located interface for social competence training of children with autism spectrum disorder. In *Human-Computer Interaction–INTERACT 2011* (pp. 123–140). Springer Berlin Heidelberg. DOI: 10.1007/978-3-642-23771-3_11.

Zancanaro, M., Pianesi F., Stock O., Venuti P., Cappelletti A., Iandolo G., Prete M., and Rossi F. (2006). Children in the museum: an environment for collaborative storytelling. Stock O., Zancanaro M. (Eds.) *PEACH: Intelligent Interfaces for Museum Visits. Cognitive Technologies Series*, Springer, Berlin. DOI: 10.1007/3-540-68755-6_8. 77, 79

Zarin, R. and Fallman, D. (2011). Through the troll forest: exploring tabletop interaction design for children with special cognitive needs. In *Proceedings of the SIGCHI Conference on Human Factors in Computing Systems* (pp. 3319–3322). ACM. DOI: 10.1145/1978942.1979434. 57, 77, 78

Zhao, H., Swanson, A. R., Weitlauf, A. S., Warren, Z. E., and Sarkar, N. (2018a). Hand-in-hand: A communication-enhancement collaborative virtual reality system for promoting social interaction in children with autism spectrum disorders. *IEEE Transactions on Human-Machine Systems*, 48(2), 136–148. DOI: 10.1109/THMS.2018.2791562. 89

Zhao, H., Zheng, Z., Swanson, A., Weitlauf, A., Warren, Z., and Sarkar, N. (2018b). Design of a Haptic-Gripper virtual reality system (Hg) for analyzing fine motor behaviors in children with autism. *ACM Transactions on Accessible Computing* (TACCESS), 11(4), 19. DOI: 10.1145/3231938. 89

Zhu, R., Hardy, D., and Myers, T. (2019). Co-designing with adolescents with autism spectrum disorder: From ideation to implementation. In *Proceedings of the 31st Australian Conference on Human-Computer-Interaction* (pp. 106–116). DOI: 10.1145/3369457.3370914. 12

Zola, K. I. (1983). Developing new self-images and interdependence. In N. Crewe, I. K. Zola, and &. Associates (Eds.), *Independent Living for Physically Disabled People* (pp. 49–59). San Francisco, Washington, London: Jossey-Bass. DOI: 10.1111/j.1468-0009.2005.00436.x. 10

Zwaigenbaum, L., Thurm, A., Stone, W., Baranek, G., Bryson, S., Iverson, J., Kau, A., Klin, A., Lord, C., Landa, R., Rogers, S., and Sigman, M. (2007). Studying the emergence of autism spectrum disorders in high-risk infants: Methodological and practical issues. *Journal of Autism and Developmental Disorders*, 37:466, 480. DOI: 10.1007/s10803-006-0179-x. 109

Authors' Biographies

Dr. Julie A. Kientz is a Professor at the University of Washington in the department of Human Centered Design and Engineering, with adjunct appointments in Computer Science and the Information School. She has worked in the space of autism and technology for the last 15 years, as well as the more general area of technologies for health, education, and families. Her background is in Computer Science, and thus she comes to this area from the perspective of a technologist, but she has had a focus in human-centered design and works to bring the perspective of end users and other stakeholders in the design of novel technologies. Her primary experience in this area has been in the development and evaluation of four technologies for individuals with autism and their caregivers. The first, Abaris, was a tool that used digital pen technology and voice recognition to help therapists and teachers conducting discrete trial training therapy become more efficient and reflective of the data they collect. The second, Baby Steps, is a long-term project looking at using a variety of software, Web, mobile, and social media technologies to engage parents of young children to identify early warning signs of developmental delay, including autism. The most recent project, led by her former Ph.D. student Kiley Sobel, was Incloodle, a shared tablet-based picture-taking application for kindergarteners to promote inclusive play and teach social-emotional understanding of neurodiversity between neurodiverse children and their peers. Dr. Kientz received a National Science Foundation CAREER award for her work on using technology to track developmental milestones in young children and was named an MIT Technology Review Top Innovator Under 35. She received her Ph.D. in Computer Science from the Georgia Institute of Technology in 2008.

Dr. Gillian R. Hayes is the Robert A. and Barbara L. Kleist Professor of Informatics at the University of California, Irvine, in the Department of Informatics in the School of Information and Computer Sciences, in the Department of Pediatrics in the School of Medicine, and in the School of Education. She is the Vice Provost for Graduate Education and Dean of the Graduate Division at UC Irvine. She is an alumna of Vanderbilt University (B.S., 1999) and the Georgia Institute of Technology (Ph.D., 2007). For nearly two decades, her research has focused on designing, developing, and evaluating technologies in support of vulnerable populations, including those with autism. Building on a background in computer science and a consulting career before academia, she focuses on methods for including people not traditionally represented in the design process or in research. She received a CAREER award from the National Science Foundation in 2008 for her work on mobile technologies for children and families coping with chronic illness and neurodevelopmental disabilities. Her most recent work has focused on both augmented reality and virtual worlds in collaboration with former students and post-doctoral scholars, Dr. LouAnne Boyd, Dr. Franceli Cibriani, Dr. Kathryn Ringland, and Dr. Monica Tentori. She has had the privelege of working with a variety of students and researchers with disabilities. She is also the co-founder of Tiwahe Technology, a technology services firm focused on classroom-based and transition technologies for schools. Following in the footsteps of Dr. Abowd, co-author on this book, Dr. Hayes received the CHI Social Impact Award in 2019 for her work supporting community-based engaged research, including work with partners in autism research and treatment.

Dr. Matthew S. Goodwin is an Interdisciplinary Associate Professor with tenure at Northeastern University jointly appointed in the Bouvé College of Health Sciences and the Khoury College of Computer and Information Sciences, where he is a founding member of a new doctoral program in Personal Health Informatics and directs the Computational Behavioral Science Laboratory. Goodwin is also a Visiting Associate Professor in the Department of Biomedical Informatics at Harvard Medical School and was previously an Adjunct Associate Professor of Psychiatry and Human Behavior at Brown University (2008–2018) and Director of Clinical Research at the MIT Media Lab (2008–2011). He has previously served on the Executive Board of the International Society for Autism Research and the Scientific Advisory Board for Autism Speaks. He has over 25 years of research and clinical experience working with children and adults on the autism spectrum and developing and evaluating innovative technologies for behavioral assessment and intervention, including video and audio capture, telemetric physiological monitors, accelerometry sensors, and digital video/facial recognition systems. Goodwin has received several honors, including a dissertation award from the Society of Multivariate Experimental Psychology, Peter Merenda Prize in Statistics and Research Methodology from the University of Rhode Island, Hariri Award for Transformative Computational Science, named an Aspen Ideas Scholar by the Aspen Institute, and a career contribution award from the Princeton Autism Lecture Series. He has obtained research funding from a variety of sources, including the National Institutes of Health, National Science Foundation, National Endowment for the Arts, Department of Defense, Simons Foundation, Nancy Lurie Marks Family Foundation, and Autism Speaks. Goodwin received his B.A. in psychology from Wheaton College and his M.A. and Ph.D., both in experimental psychology and behavioral science, from the University of Rhode Island. He completed a postdoctoral fellowship in Affective Computing in the MIT Media Lab in 2010.

Dr. Mirko Gelsomini is a Postdoc at the Department of Electronics, Information and Bioengineering (DEIB) and an Adjunct Professor at the Department of Design (DESIGN) of Politecnico di Milano. He received his Ph.D. with honors (2018) in Information Technology and holds a master degree with honors (2014) in Engineering of Computing Systems at Politecnico di Milano. His research focuses on the analysis, design, development, and evaluation of innovative interactive technologies, such as virtual reality, social robots, multisensory environments and motion-based interaction, to support play, learning, and inclusion of children with neuro-developmental disorders. Winner of different international prizes and author of more than 50 publications on the topic, he worked at Georgia Institute of Technology in the Ubicomp Group and at Massachusetts Institute of Technology (Media Lab) in the Personal Robots Group.

Dr. Gregory D. Abowd is a Regents' and Distinguished Professor in the School of Interactive Computing at the Georgia Institute of Technology. He is the father of two boys, Aidan and Blaise, who have been diagnosed on the autism spectrum. Since the early 2000s, he has devoted a large portion of his research career to developing technologies addressing challenges related to autism. He advised, and was subsequently inspired by, the doctoral research of Gillian Hayes and Julie Kientz, two of the coauthors of this book, and has advised numerous doctoral students on topics in this area, ranging from direct interventions to tools for clinicians, educators, or researchers to use in screening, diagnosis, and assessment of interventions. He is the Chief Research Officer of Behavior Imaging Solutions, which has commercialized some of the thesis research of the CareLog system designed and evaluated by Gillian Hayes and is currently pursuing commercialization of a portable in-home behavior capture system that is the thesis research of current Ph.D. student Nazneen. Gregory served on the Innovative Technologies for Autism Committee with Matthew Goodwin that was first part of the Cure Autism Now Foundation and has continued under the auspices of Autism Speaks. In 1998, he founded the Atlanta Autism Consortium to unite different stakeholder communities within the Atlanta area focused on research, education, and advocacy, and he now serves as the president of that non-profit organization. He has published extensively in the area of technology and autism and has received several professional awards from the Association of Computing Machinery (ACM) in recognition of that work, including being selected as a Fellow of the ACM.

Printed in the United States
by Baker & Taylor Publisher Services